Solving Math Word Problems

Solving Math Word Problems

Teaching Students with Learning Disabilities Using Schema-Based Instruction

Asha K. Jitendra

8700 Shoal Creek Boulevard
Austin, Texas 78757-6897
800/897-3202   Fax 800/397-7633
www.proedinc.com

© 2007 by PRO-ED, Inc.
8700 Shoal Creek Boulevard
Austin, Texas 78757-6897
800/897-3202   Fax 800/397-7633
www.proedinc.com

All rights reserved. No part of the material protected by this copyright notice may be reproduced or used in any form or by any means, electronic or mechanical, including photocopying, recording, or by any information storage and retrieval system, without prior written permission of the copyright owner.

NOTICE: PRO-ED grants permission to the user of this material to make unlimited copies of the pages on the CD-ROM for teaching or clinical purposes. Duplication of this material for commercial use is prohibited.

Note. The Change and Vary diagrams are adapted from *Schemas in Problem Solving* (p. 135), by S. P. Marshall, 1995, New York: Cambridge University Press. Copyright 1995 by Cambridge University Press. Adapted with permission.

Library of Congress Cataloging-in-Publication Data

Jitendra, Asha K.
   Solving math word problems : teaching students with learning disabilities using schema-based instruction / Asha K. Jitendra.
      p. cm.
   ISBN-13: 978-1-4164-0245-9
   ISBN-10: 1-4164-0245-4
   1. Arithmetic—Problems, exercises, etc.   2. Mathematics—Problems, exercises, etc.   3. Mathematics—Study and teaching (Elementary)   4. Mathematics—Study and teaching (Middle school)   5. Learning disabled children—Education.   I. Title.
   QA139.J57 2007
   513.076—dc22

                                                                                2006031916

Art Director: Jason Crosier
Designer: Sandy Salinas
This book is designed in Lucida Sans and Lucida Sans Narrow.

Printed in the United States of America

1   2   3   4   5   6   7   8   9   10   11 10 09 08 07

Contents

Acknowledgment vii

Introduction to the Program ix

PART I

Addition and Subtraction   1
(Asha K. Jitendra)

Unit 1: Change Problems   3
Lesson 1   3

Lesson 2   15

Lesson 3   27

Lesson 4   35

Lesson 5   41

Unit 2: Group Problems   45
Lesson 6   45

Lesson 7   57

Lesson 8   69

Lesson 9   77

Lesson 10   81

Unit 3: Compare Problems   85
Lesson 11   85

Lesson 12   97

Lesson 13   109

Lesson 14   117

Lesson 15   121

Unit 4: One-Step Problem Review: Change, Group, and Compare Problems   125
Lesson 16   125

Lesson 17   135

Lesson 18   139

Unit 5: Two-Step Problems and Mixed Review: Change, Group, and Compare Problems   143

Lesson 19   143

Lesson 20   155

Lesson 21   159

PART II

Multiplication and Division   169
  (Yan Ping Xin and Asha K. Jitendra)

Unit 1: Multiplicative Compare Problems   171

Lesson 1   171

Lesson 2   185

Lesson 3   199

Lesson 4   211

Lesson 5   217

Unit 2: Vary Problems   219

Lesson 6   219

Lesson 7   233

Lesson 8   247

Lesson 9   259

Lesson 10   265

Unit 3: Multiplicative Compare and Vary Problem Review   269

Lesson 11   269

Lesson 12   275

Lesson 13   279

About the Author   283

Acknowledgment

I would like to thank Yan Ping Xin, the primary author of the strand on solving multiplication and division problems in this text. The preparation of the teaching and student materials related to the Multiplicative Compare Problems and Vary Problems sections in *Solving Math Word Problems* is based on her dissertation work and provides empirical support for schema-based instruction in enhancing the word problem–solving performance of students with learning difficulties.

Introduction to the Program

Solving Math Word Problems: Teaching Students with Learning Disabilities Using Schema-Based Instruction is a teacher-directed program designed to teach critical word problem–solving skills to students with disabilities in the elementary and middle grades. The program is carefully designed to promote conceptual understanding using schema-based instruction (SBI) and provides the necessary scaffolding to support learners who struggle with math word problems. The program features are consistent with national and statewide school reform movements toward challenging academic standards and with the regulations of the No Child Left Behind Act of 2001 (NCLB) and the Individuals with Disabilities Education Improvement Act of 2004 (IDEIA). For example, IDEIA requires access to the general education curriculum for students with learning disabilities. This means ensuring a shift in focus from rote memorization of mathematical procedures to conceptual understanding of mathematical concepts, skills, and relations.

Solving Math Word Problems is organized around two strands: (a) solving addition and subtraction problems and (b) solving multiplication and division problems. The program includes the following features.

- The program focuses on the "big ideas" or salient problem schemata involved in solving addition, subtraction, multiplication, and division word problems. The problem schemata that pertain to a wide range of problems involving all four operations include Change, Group, Compare (additive), Multiplicative Compare, and Vary (Marshall, Pribe, & Smith, 1987). A schema as a knowledge structure serves the function of knowledge organization. According to Marshall (1995), schemata are the basis for understanding and the appropriate mechanism for the problem solver to "capture both the patterns of relationships as well as their linkages to operations" (p. 67). These problem types characterize word problems typically found in elementary and middle grades and are the emphasis of this program (Van de Walle, 1998).
- The program features tasks based on word problems presented in commonly adopted U.S. mathematics textbooks. Word problems are varied and are formatted as text, graphs, tables, and pictographs.
- The program has instruction addressing both conceptual knowledge and procedural knowledge, which are critical to

successful mathematical problem solving (e.g., Hegarty, Mayer, & Monk, 1995). Conceptual knowledge requires problem comprehension and representation, which involve translating the text of the problem into a semantic representation, based on an understanding of the problem type. Whereas procedural knowledge of operations is important, instruction should facilitate "a highly integrated understanding of the operations and the many different but related meanings these operations take on in real contexts" (Van de Walle, 1998, p. 117). The big ideas for developing meanings for the operations should, for example, show that addition and subtraction are connected. In *Solving Math Word Problems,* for example, SBI uses schemata diagrams to represent the information in word problems and to help students figure out what operation is needed to solve the problem (Van de Walle, 1998).

- The program provides appropriate scaffolding of instruction, including the following:
  1. Teacher-mediated instruction is followed by paired-partner learning and independent learning activities.
  2. The first lesson in each unit contains story situations, and subsequent lessons in each unit contain word problems with unknown information.
  3. Initially, diagrams and checklists are provided to support student learning. Later, students construct their own diagrams.
- The program provides adequate practice and a mixed review of problem types.
- Instruction is aligned with national standards in terms of problem solving, communicating, connecting, reasoning, and representing word problems.
- Progress-assessment measures are provided to monitor students' progress in solving word problems.

Program Components

The program is divided into units and lessons. Each unit introduces a problem type—that is, change, group, compare, multiplicative compare, or vary—and the first lesson in each unit contains problem schema instruction for that problem type. Subsequent lessons within each unit focus on problem solution instruction for that problem type. Each lesson begins with a list of materials needed for that lesson (provided as printable forms on the accompanying CD-ROM), a teacher-scripted procedure (contained in this manual), and answers to the worked-out problems (see the Answer Sheets and Reference Guides on the CD-ROM).

During the *problem schema instruction phase* (i.e., the first lesson in each unit), students are provided with *story* situations that contain only known information. They are taught to identify the problem schema (e.g., change, group, compare) and to represent the features of the story situation using schematic diagrams. The aim of this phase is to show students how to understand the underlying structure of the problem type. Students first learn to interpret and elaborate on the main features of the story situation. Next students map the details of the story onto the schema diagram. This step ensures that all irrelevant information in the story is discarded and that problem representation is based on schema elaboration knowledge.

During the *problem solution phase* (i.e., subsequent lessons in each unit), students learn to solve problems with unknowns. A four-step strategy checklist with the acronym FOPS—Find the problem type, Organize the information in the problem using the diagram, Plan to solve the problem, and Solve the problem—is used to anchor students' learning of the word problem–solving strategy to solve word problems (see the Checklists folder on the CD-ROM). Eventually, schematic diagrams are systematically faded at the end of the instructional unit on each problem type.

Following is a list of the printable forms that are provided on the accompanying CD-ROM to support the teaching of the program:

- *Diagrams* for all problem types (use as write-on transparencies; make enlarged copies and either laminate and use as write-on posters or copy on card stock for classroom display)
- *Checklists* for all story and problem types (use as write-on transparencies; make enlarged copies and either laminate and use as write-on posters or copy on card stock for classroom display; pass out individual laminated copies for students)
- *Overhead Modeling* of stories and problems (use as write-on transparencies; pages contain diagrams and space to write out, explain, and model problem-solving processes on the overhead)
- *Reference Guides* for particular lessons (use as transparencies to efficiently illustrate how the problem was solved following a verbal explanation and discussion)
- *Answer Sheets* (make copies and pass out to students to use during paired learning; students can use sheets to correct their errors)
- *Student Pages* (make copies of worksheets that students are required to complete for lessons)
- *Progress Assessment* (print and pass out to students; samples of math word problem–solving forms for monitoring students' progress; results from these assessments can be used to inform instruction and practice)

Using the Program

Although a scripted, detailed teaching procedure is provided in this manual to ensure consistency in implementing the critical content, we recommend that you use the scripts only as a framework for instructional implementation. We do not recommend reading the script verbatim but rather suggest becoming familiar with the script and then using your own explanations and elaborations to implement SBI.

The program can be used whenever students are to solve arithmetic word problems that involve addition, subtraction, multiplication, or division. It can be used in varied settings (general education programs, Title I programs, special education programs) and is designed for flexible use with children of varied needs, primarily those who are at risk for mathematics failure or who may have learning, attention, organizational, and memory difficulties. It can be implemented with individual students or during small- and whole-group instruction. Consider the following conditions, however, when implementing the program:

▶ *Are students exposed to several problem-solving strategies (e.g., working backwards, using a model, guess and check) at the same time?* If so, the benefits of SBI may be compromised for students with disabilities, who may experience cognitive information overload (see Jitendra, DiPipi, & Grasso, 2001).

▶ *What is the difficulty level of the word problems when students are first introduced to SBI?* If the problems are too difficult for students, their ability to understand and map the information onto the schematic diagram could be undermined. The goal is for students to learn how to use the strategy; therefore, initial problems should be ones that students are able to read and understand.

▶ *When cooperative learning groups are employed during SBI, have students received sufficient time to master the new material individually?* This is important to prevent students with disabilities from assuming a passive role in the group (see Jitendra et al., 2001).

▶ *When implementing SBI in general education classrooms, do some students need more intense and systematic instruction than others?* Teachers should consider the importance of appropriately mediating instruction (e.g., providing extended practice, additional explanations and elaborations) for students with disabilities to be successful problem solvers (see Jitendra et al., 2001; Jitendra, Griffin, Deatline-Buchman, & Sczesniak, in press).

Guide to Paired Learning

▶ *Assign partners before the assignment.* By assigning partners before the activity, instructional time loss is minimized. Partners should be heterogeneously grouped by achievement. For example, rank-order individuals in the group according to their math performance, and then divide the group in half. Pair the top performer in Group 1 with the top performer in Group 2.

▶ *Change partners.* By changing partners, students get the opportunity to work with other students in the class.

▶ *Monitor discussions.* It is important that the students work in an area that you can easily access. By circulating around the classroom and monitoring student work, you will be able to assess student comprehension of the word problem–solving task. This can be beneficial for deciding when to reteach specific information or whether to provide immediate remediation. In addition, if one student in a pair appears to be doing all the work, you can address the situation accordingly.

▶ Teach students to use a think–plan–share procedure.

Think: Have each student in the pair independently read the problem and think about the features of the word problem to figure out the problem type.
Plan: Have each student plan to solve the problem by organizing the information using the given diagram and then solve it.
Share: Ask students to share their plans and answers with their assigned partners. Students with different answers should discuss how each solved the problem and correct their errors using the Answer Sheet. This may be a good time for one student to role-play the teacher and the other the student.

Program Audience

The addition and subtraction word problem–solving lessons are designed for third graders but can be used with second graders by modifying the difficulty level of the language and computation skills. In addition, the lessons can be used with older children who have experienced consistent difficulties in solving addition and subtraction word problems.

The multiplication and division problems are appropriate for middle school students (e.g., fifth to eighth graders). This program can serve as a supplement to word problem–solving instruction presented in published mathematics textbooks.

Solving Math Word Problems is designed primarily for school practitioners (e.g., special education and general education teachers, school psychologists, supervisors). In addition, the program is useful for teacher educators during preservice training (e.g., for undergraduate- and graduate-level teaching methods courses) or in-service training (e.g., as a desk reference for professionals).

Lesson Time Frames

Each lesson is designed to require about 50 to 60 minutes. Some lessons (e.g., mixed review of one-step problems) are shorter (about 30 minutes). If a lesson or problem is not completed in one class period, it can be completed on the following day.

Judging Program Effectiveness

An effective way to determine whether or not SBI is working is to frequently question and evaluate students' performance at the end of each lesson. This evaluation serves to check students' problem-solving knowledge and determine whether they mastered the strategy steps. When students incorrectly respond to a word problem, review their work to examine possible errors related to strategy use. This information can be used to provide corrective feedback and make modifications (e.g., by modeling more examples or rephrasing the word problem). Examine students' independent worksheets for strategy use (e.g., drawing a diagram, mapping information onto the diagram, planning, writing the number sentence) and provide them with additional instruction as needed before moving to the next problem type. Over time and with frequent practice, students should be able to explain the features of the problem types and verbalize the strategy steps as they solve different problems.

Administer a progress assessment measure (see accompanying CD-ROM) once every 1 to 2 weeks to monitor student performance. In addition, evaluate students' maintenance of strategy use over time and transfer of problem-solving skills to solve novel and complex problems. Finally, assess student satisfaction to determine the benefits of SBI. If SBI is not having the desired effects, ask yourself the following questions:

- Has the student mastered the prerequisite skills (e.g., identifying the different problem schemata) to a criterion level?
- Has sufficient modeling of strategy steps using several examples and explanations been provided?
- Has systematic and varied practice been provided?

What You Need To Know About the Program

The program requires comprehensive, teacher-directed instruction to promote student success. It also depends on teacher orchestration of classroom management skills, delivery of instruction, and fidelity to the program. Although instruction should be explicit, it is equally necessary that you employ frequent student exchanges (opportunities for student responses) to facilitate the identification of critical elements of the problem schema. As such, checking student understanding and providing appropriate feedback on strategy usage on an ongoing basis is essential. For example, when students inaccurately identify the problem type, remind them to check the strategy steps and apply them in the correct sequence.

As with any instructional program, time is necessary for the program to be effective for students with learning problems. The more instruction and practice students receive, the more likely they are to make strong progress. Scheduling sufficient review time is necessary to facilitate acquisition and maintenance of the taught skill.

The program also requires differentiation of instruction based on student ability levels; low-performing students, in particular, may need more instructional support (e.g., explicit instruction, diagrams, checklists) than others to reach their potential. Instruction should be criterion based rather than time based for these students. Ensure that students are proficient in verbalizing the strategy steps and solving problems using schemata diagrams prior to removing the checklists and diagrams. All students need to be exposed to a variety of problems to promote generalization of the problem-solving skill.

Research Evidence To Support the Program

A growing body of literature on mathematical problem solving provides empirical support for SBI (e.g., Jitendra & Xin, 1997; Xin & Jitendra, 1999). SBI is known to benefit elementary, middle, and high school students with learning disabilities and students at risk for math failure (Hutchinson, 1993; Jitendra et al., 1998; Jitendra & Hoff, 1996; Jitendra, Hoff, & Beck, 1999; Zawaiza & Gerber, 1993), as well as students without disabilities (e.g., Fuchs, Fuchs, Finelli, Courey, & Hamlett, 2004; Fuchs et al., 2003a; Fuchs et al., 2003b; Fuchs, Fuchs, Prentice, et al., 2004; Jitendra et al., 2007; Jitendra et al., in press). The following studies represent a decade of research that has examined the effectiveness of SBI for enhancing students' mathematical problem-solving skills: Jitendra et al. (2001); Jitendra, DiPipi, and Perron-Jones (2002); Jitendra et al. (in press); Jitendra et al. (2007); Jitendra et al. (1998); Jitendra and Hoff (1996); Jitendra et al. (1999); and Xin, Jitendra, and Deatline-Buchman (2005).

References

Fuchs, L. S., Fuchs, D., Finelli, R., Courey, S. J., & Hamlett, C. L. (2004). Expanding schema-based transfer instruction to help third graders solve real-life mathematical problems. *American Educational Research Journal, 41,* 419–445.

Fuchs, L. S., Fuchs, D., Prentice, K., Burch, M., Hamlett, C. L., Owen, R., et al. (2003a). Explicitly teaching for transfer: Effects on third-grade students' mathematical problem solving. *Journal of Educational Psychology, 95,* 293–305.

Fuchs, L. S., Fuchs, D., Prentice, K., Burch, M., Hamlett, C. L., Owen, R., et al. (2003b). Enhancing third-grade students' mathematical problem solving with self-regulated learning strategies. *Journal of Educational Psychology, 95,* 306–315.

Fuchs, L. S., Fuchs, D., Prentice, K., Hamlett, C. L., Finelli, R., & Courey, S. J. (2004). Enhancing mathematical problem solving among third-grade students with schema-based instruction. *Journal of Educational Psychology, 96,* 635–647.

Hegarty, M., Mayer, R. E., & Monk, C. A. (1995). Comprehension of arithmetic word problems: A comparison of successful and unsuccessful problem solvers. *Journal of Education Psychology, 87,* 18–32.

Hutchinson, N. L. (1993). Effects of cognitive strategy instruction on algebra problem solving of adolescents with learning disabilities. *Learning Disabilities Quarterly, 16,* 34–63.

Individuals with Disabilities Education Improvement Act of 2004, 20 U.S.C. § 1400 *et seq.*

Jitendra, A. K., DiPipi, C. M., & Grasso, E. (2001). The role of a graphic representational technique on the mathematical problem solving performance of fourth graders: An exploratory study. *Australasian Journal of Special Education, 25,* 17–33.

Jitendra, A. K., DiPipi, C. M., & Perron-Jones, N. (2002). An exploratory study of word problem–solving instruction for middle school students with learning disabilities: An emphasis on conceptual and procedural understanding. *Journal of Special Education, 36,* 23–38.

Jitendra, A. K., Griffin, C., Deatline-Buchman, A., & Sczesniak, E. (in press). Mathematical word problem-solving in third grade classrooms: Lessons learned from design experiments. *Journal of Educational Research.*

Jitendra, A. K., Griffin, C., Haria, P., Leh, J., Adams, A., & Kaduvetoor, A. (2007). A comparison of single and multiple strategy instruction on third grade students' mathematical problem solving. *Journal of Educational Psychology, 99,* 115–127.

Jitendra, A. K., Griffin, C., McGoey, K., Gardill, C., Bhat, P., & Riley, T. (1998). Effects of mathematical word problem solving by students at risk or with mild disabilities. *Journal of Educational Research, 91,* 345–356.

Jitendra, A. K., & Hoff, K. (1996). The effects of schema-based instruction on mathematical word problem solving performance of students with learning disabilities. *Journal of Learning Disabilities, 29,* 422–431.

Jitendra, A. K., Hoff, K., & Beck, M. (1999). Teaching middle school students with learning disabilities to solve multistep word problems using a schema-based approach. *Remedial and Special Education, 20,* 50–64.

Jitendra, A. K., Sczesniak, E., & Deatline-Buchman, A. (2005). Validation of curriculum-based mathematical word problem solving tasks as indicators of mathematics proficiency for third graders. *School Psychology Review, 34,* 358–371.

Jitendra, A. K., & Xin, Y. P. (1997). Mathematical word problem solving instruction for students with disabilities and at risk: A research synthesis. *Journal of Special Education, 30,* 412–439.

Marshall, S. P. (1995). *Schemas in problem solving.* New York: Cambridge University Press.

Marshall, S. P., Pribe, C. A., & Smith, J. D. (1987). *Schema knowledge structures for representing and understanding arithmetic story problems* (Tech. Rep. Contract No. N00014-85-K-0061). Arlington, VA: Office of Naval Research.

No Child Left Behind Act of 2001, 20 U.S.C. 70 § 6301 *et seq.*

Van de Walle, J. A. (1998). *Elementary and middle school mathematics: Teaching developmentally* (3rd ed.). Boston: Allyn & Bacon.

Xin, Y. P., & Jitendra, A. K. (1999). The effects of instruction in solving mathematical word problems for students with learning problems: A meta-analysis. *The Journal of Special Education, 32,* 207–225.

Xin, Y. P., Jitendra, A. K., & Deatline-Buchman, A. (2005). Effects of mathematical word problem solving instruction on students with learning problems. *Journal of Special Education, 39,* 181–192.

Zawaiza, T. B. W., & Gerber, M. M. (1993). Effects of explicit instruction on community college students with learning disabilities. *Learning Disabilities Quarterly, 16,* 64–79.

PART I

Addition and Subtraction

Unit 1

Change Problems

Lesson 1: Problem Schema

Materials Needed

Checklist	Change Story Checklist (laminated copies for students and poster or transparency for display)
Diagram	Change Problem diagram poster
Overhead Modeling	Lesson 1: Change Stories 1, 2, and 3
Student Pages	Lesson 1: Change Stories 1, 2, and 3
	Lesson 1: Change Schema Worksheet 1

Teacher: You will learn to solve word problems, which can help you use math in everyday life. For example, when you shop, you can figure out the costs of items and decide whether you have enough money to buy them. Solving word problems can help you improve your math and help you understand how math applies to everyday life. You already know how to add and subtract. You will learn to use these operations to solve word problems. Today you will learn a type of addition/subtraction problem called "change." You will learn how to organize information in change stories by using diagrams. Later you will learn to solve change word problems. In a change problem, there is a beginning amount (e.g., "John had 6 cookies"). Then a change action (e.g., "John got 2 more cookies") occurs that increases the beginning amount. Finally we end up with a different amount (i.e., 8 cookies) from when we started. Sometimes the change action (e.g., "John ate 2 cookies") decreases the beginning amount. In this situation the ending amount (4 cookies) is less than the beginning amount.

Change Story 1

Teacher: (*Display Change Story Checklist.*) Here are two steps we will use to organize information in a change story. (*Point to each step on Change Story Checklist and read each one.*) Let's use these two steps to do an example. Look at this story. (*Display Overhead Modeling page for Change Story 1. Pass out student copies of Change Story 1 and laminated* copies of Change Story Checklist. Point to first check box on Change Story Checklist.)

Now we are ready for Step 1: Find the problem type. To find the problem type, I will read the story and retell it in my own words. (*Read Change Story 1 aloud.*)

"Jane had 4 video games. Then her mother gave her 3 more video games for her birthday. Now Jane has 7 video games."

Now I'll retell the story in my own words to help me understand it. When I retell, I will ask myself, What do I know in this story? (*Retell the story.*)

Jane began with 4 video games. Then she got 3 *more* video games from her mother. She ended up with 7 video games.

I read the story and then told it in my own words. Let's check off the first box under Step 1 on the checklist.

(*Point to second check box under Step 1.*) Now I will ask myself if the story is a change problem type. This is a change story, because it has a beginning, a change, and an ending. The beginning amount of 4 video games changed when Jane's mother gave her 3 more video games. This change action (gave more) *increased* the number of video games so that Jane ended up with 7 video games. Also, the beginning, change, and ending in a change story all describe (or talk about) the same thing (e.g., video games). (*Check off second check box under Step 1.*)

Now I am ready for Step 2: Organize the information in the problem using the change diagram. (*Display Change Problem diagram poster.*)

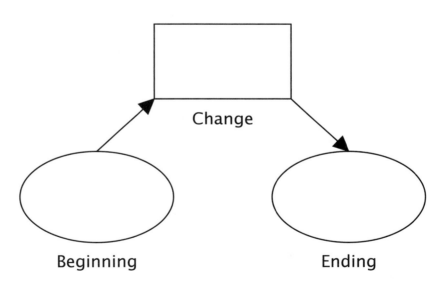

(*Point to first box under Step 2.*) To organize the information in a change problem, we first underline the label that describes (or talks about) the beginning, change, and ending. What is the label that describes the beginning, change, and ending in this story?

Students: Video games.

Teacher: Good, let's underline video games in the story and write "video games" in the diagram for the beginning, change, and ending. (*Pause for students to complete.*) Let's check off the first box under Step 2 on the checklist.

(*Point to second box under Step 2.*) Now let's read the story to underline and circle information about the beginning, change, and ending. The first sentence says, "Jane had 4 video games." This sentence tells us about the beginning amount. How do I know? The word had (*underline*) in this sentence tells us that Jane began with 4 video games. Do you know the beginning amount from this sentence?

Students: Yes, 4 video games.

Teacher: Circle "4" and write it in the diagram for the beginning amount. The next sentence says, "Then her mother gave her 3 more (*underline*) video games for her birthday." This sentence describes the change in this story. That is, the change action ("gave more") increased the beginning amount. What is the change amount in this story?

Students: 3.

Teacher: Circle "3" and write "+ 3" (to indicate getting more) for the change amount in the diagram. (*Pause for students to complete.*) The last sentence says, "Now Jane has 7 video games." The words now has (*underline*) tell me that Jane ended up with 7 video games. What is the ending amount in this story?

Students: 7 video games.

Teacher: Circle "7" and write it in for the ending amount in the diagram. (*Pause for students to complete.*) We underlined the important information, circled numbers, and wrote the numbers in the change diagram. Check off the second box under Step 2 on the checklist. Now let's look at the diagram and read what it says.

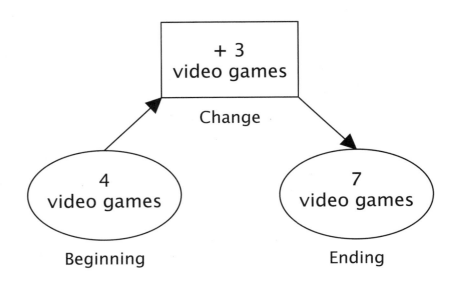

(*Point to relevant parts of the diagram as you explain.*) We have a beginning amount of 4 video games, a change of 3 more video games, and an ending amount of 7 video games. In this change story, the beginning, change, and ending all talk about the same thing (i.e., video games). Does this make sense that we began with 4 video games and ended with 7 video games? Why?

Students: Yes, because the change action (i.e., gave more) in this story caused an increase to the beginning amount. Therefore, the ending amount is greater than the beginning amount.

Teacher: That is right. The change of 3 *more* video games increased the beginning amount so that Jane ended with more video games (7) than she started with (4). In this change story, the ending amount is the total or whole. Let's review this change problem: (1) There is a beginning, a change, and an ending (*point to diagram*), and (2) we began with video games and ended with video games. The change also involved video games. (*Point to diagram.*) What's this problem called and why?

Students: Change, because it has a beginning, a change, and an ending amount.

Teacher: Excellent. Also, the beginning, change, and ending all describe video games in this story.

Change Story 2

Teacher: (*Display Overhead Modeling page of Change Story 2. Have students look at student copies of Change Story 2.*)
Touch Story 2. (*Point to Change Story Checklist.*) What's the first step?

Students: Find the problem type.

Teacher: Right. (*Point to first check box on Change Story Checklist.*) To find the problem type, I will read the story and retell it in my own words. (*Read story aloud.*)

"Before he gave away 14 marbles, James had 36 marbles. Now he has 22 marbles."

I read the story. What must I do next?

Students: Retell the story using own words.

Teacher: Yes, I will retell the story in my own words to help me understand it. When I retell, I will ask myself, What do I know in this story? (*Retell story aloud.*)

James began with 36 marbles. Then he gave away 14 marbles. He now has 22 marbles left.

I read the story and then told it in my own words. Let's check off the first box under Step 1 on the checklist. (*Point to second check box under Step 1.*) Now I will ask myself if the story is a change problem. Why do you think this is a change story? What does the story describe?

Students: It has a beginning, a change, and an ending. They all describe marbles.

Teacher: Good. Check off the second box under Step 1. Now I am ready for Step 2: Organize the information in the story using the change diagram.

Change Problem

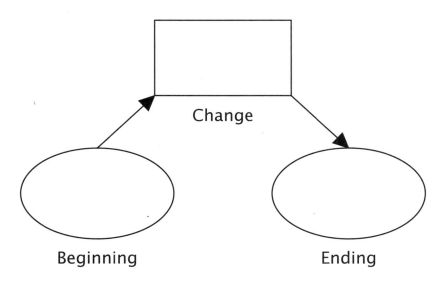

(*Point to first check box under Step 2.*) To organize the information, I need to first underline the label that describes the beginning, change, and ending. What is the label that describes the beginning, change, and ending in this story?

Students: Marbles.

Teacher: Good, we will underline marbles and write it in the diagram for the beginning, change, and ending. (*Pause for students to complete.*) Let's check off the first box under Step 2 on the checklist. (*Point to second box under Step 2.*) Now let's read the story to underline information about the beginning, change, and ending.

The first sentence says, "Before he gave away 14 marbles, James had 36 marbles."

(*Hint: Indicate to students that the first sentence does not always start with the beginning amount and that we have to read the problem carefully to find the beginning, change, and ending amounts.*)

This sentence tells us about the change and beginning amounts. The word had (*underline*) in this sentence tells us that James began with 36 marbles. What is the beginning amount?

Students: 36 marbles.

Teacher: Right. Circle "36" and write it in for the beginning amount. (*Pause for students to complete.*) This sentence also says that he gave away 14 marbles. The words gave away (*underline*) tell about a change action. Does the change in this story increase or decrease the beginning amount. How do you know?

Students: Decrease, because when you give something away, you have less of it.

Teacher: That's right! What is the change amount?

Students: 14.

Teacher: Circle 14 and write in "− 14" (to indicate "less" or a decrease) for the change amount. (*Pause for students to complete.*) The last sentence says, "Now he has 22 marbles." The words now has (*underline*) tell me that James ended up with 22 marbles. What is the ending amount in this story?

Students: 22.

Teacher: Circle "22" and write it in for the ending amount in the diagram. (*Pause for students to complete.*)

We underlined and circled the important information and wrote the numbers in the change diagram. Let's check off the second box under Step 2 on the checklist. Now look at the diagram and read what it says.

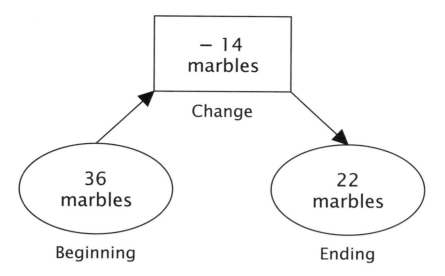

Students: There was a beginning amount of 36 marbles. Then there was a change of 14 less marbles. Now there are 22 marbles.

Teacher: In this change story, the beginning, change, and ending amounts all talk about the same thing (i.e., marbles). Does this make sense that we began with 36 marbles and ended up with 22 marbles? Why?

Students: Yes, because the change action (i.e., gave away) in this story caused a decrease to the beginning amount. Therefore, the ending amount is less than the beginning amount.

Teacher: That's right. The change (14 less marbles) in this story decreased the beginning amount and James ended with less marbles (22) than he started with (36). In this change story, the beginning amount is the total or whole.

Let's review this change problem: (1) There is a beginning, a change, and an ending (*point to the diagram*), and (2) we began with marbles and ended with marbles. The change also involved marbles. (*Point to the diagram.*) What's this problem called? Why?

Students: Change, because it has a beginning, a change, and an ending amount.

Teacher: Excellent. Also, the beginning, change, and ending all describe marbles.

Change Story 3

Teacher: (*Use the script as a guideline for mapping information in Change Story 3, and facilitate understanding and reasoning by having frequent student–teacher exchanges. Display Overhead Modeling page of Change Story 3. Have students look at student pages of Change Story 3.*)

Touch Story 3. (*Point to first check box on Change Story Checklist.*) What's the first step? What do we need to do?

Students: Find the problem type by reading the story and retelling it.

Teacher: Great! (*Read Change Story 3 aloud or have a student read it.*)
"Tom had 42 baseball cards. He now has 55 baseball cards after he bought 13 more cards."
After reading the story, what must you do next?

Students: Retell the story using own words.

Teacher: Yes. Retell the story in your own words to help you understand it. (*Call on students to retell the story. When they retell the story, remind students to tell what they know in the story.*)

Students: Tom began with 42 baseball cards. Then he bought 13 more cards. Now he has 55 baseball cards.

Teacher: (*Check off first box under Step 1 on checklist. Point to second box under Step 1.*) What kind of a problem type is this? How do you know?

Students: It is a change problem, because the words "began with 42 baseball cards" tell about a beginning, "bought 13 more cards" tell about a change, and "now has 55 baseball cards" tell about the ending amount.

Teacher: Correct! The change action ("bought more") in this story is an increase of 13 cards. (*Check off second box under Step 1 on checklist.*)
Now you are ready for Step 2: Organize the information using the change diagram.
(*Point to first box under Step 2.*) To organize the information in a change story, what must you underline first and write in the change diagram?

Students: The label that describes the beginning, change, and ending.

Teacher: What is the label that describes the beginning, change, and ending?

Students: Baseball cards.

Teacher: Good. Underline <u>baseball cards</u> in the story and write it for the beginning, change, and ending in the diagram. (*Check off first box under Step 2 on checklist.*)
(*Point to second box under Step 2 of checklist.*) Now read each sentence to underline the important information and circle numbers in the story, and write the numbers for the beginning, change, and ending in the change diagram. What does the first sentence say?

Students: "Tom had 42 baseball cards."

Teacher: Does this sentence tell about the beginning, change, or ending? How do you know?

Students: Beginning, because the word *had* tells us that Tom began with 42 baseball cards.

Teacher: Great! Underline the word had and circle "42" in the story. Then write "42" for the beginning amount in the diagram. (*Pause for students to complete.*) What does the next sentence say?

Students: "He now has 55 baseball cards after he bought 13 more cards."

Teacher: Does this sentence tell about the change or ending? How do you know?

Students: Both ending and change, because the words "*now has* 55 baseball cards" tell about the ending amount, and the words "*bought* 13 *more* cards" tell about the change amount.

Teacher: Excellent! Underline the words now has and circle "55" in the story and write "55" in the diagram for the ending amount. (*Pause for students to complete.*) Also, underline the words bought more to indicate the change action. Does the change in this story increase or decrease the beginning amount. How do you know?

Students: Increase, because when you buy more cards, you have more of them.

Teacher: That's right! The change action (*point to* "bought more") in this story is *an increase*. What is the change amount?

Students: 13.

Teacher: Circle "13" and write in "+ 13" (to indicate "more" or an increase) for the change amount. (*Pause for students to complete.*) You underlined and circled the important information and wrote numbers in the Change Diagram. Check off the second box under Step 2 on the checklist. Now look at the diagram and read what it says.

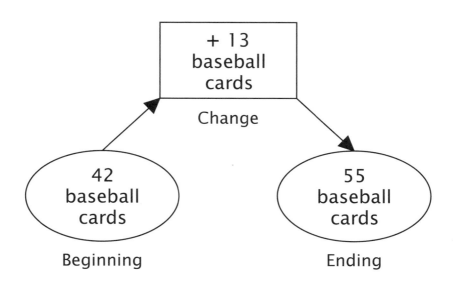

Students: There is a beginning amount of 42 baseball cards, a change amount of 13 more baseball cards, and an ending amount of 55 baseball cards.

Teacher: Does this make sense? Why?

Students: Yes, because if Tom bought 13 more baseball cards, then he should end up with more baseball cards than he started with.

Teacher: What is this problem called? Why?

Students: Change, because it has a beginning, a change, and an ending amount, which all describe baseball cards.

Teacher: That is right. The change action (i.e., bought more) increased the beginning amount so that Tom ended up with more baseball cards (55) than he started with (42). In this change story, the ending amount is the total or whole. Let's review this change story. We began with baseball cards and ended with baseball cards. The change also involved baseball cards. *(Point to diagram.)*

(Pass out Change Schema Worksheet 1.) Now I want you to do the next four stories on your own. Remember to use the two steps and organize the information in the stories using the Change Problem diagram.

(Monitor students as they work. Then check the information in the diagrams. Make sure the diagrams are labeled correctly and completely, as done below.)

Change Schema Story 1: "Linda had 47 videotapes of her favorite movies. Linda now has 26 videotapes after her brother borrowed 21 of her videotapes."

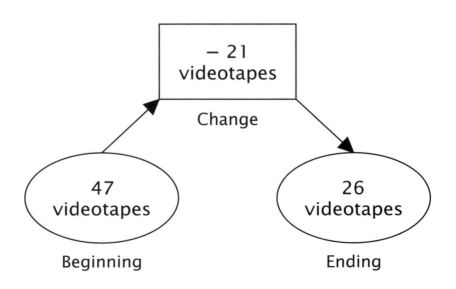

Change Schema Story 2: "Christine had 18 tropical fish before she went to the pet store and bought 12 more fish. Now Christine has 30 tropical fish."

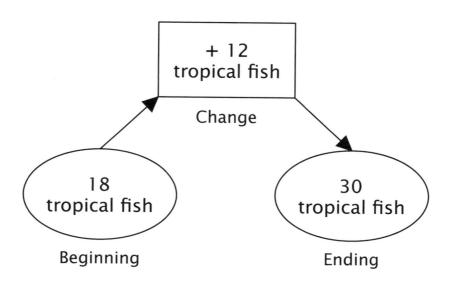

Change Schema Story 3: "Mandy began the morning by driving 23 miles. Then she drove 13 more miles. Now Mandy has driven 36 miles."

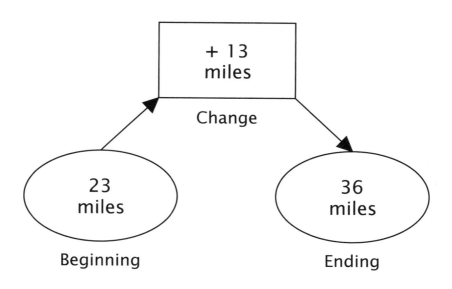

Change Schema Story 4: "4 children still remain in the tent. Before 7 children left the tent, there were 11 children in the tent."

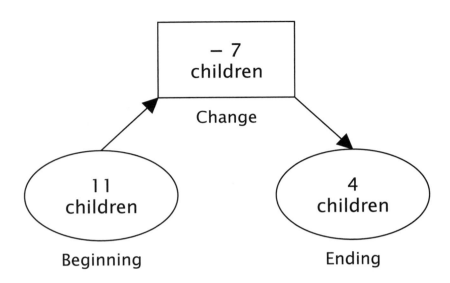

Teacher: (*Note: When checking students' work, make sure that they can tell you whether the beginning amount or the ending amount in the change story is the total or whole.*)

You learned to map information in change *stories* onto diagrams. Next you will learn to solve change *problems*. Later you will learn to organize information and solve other problem types (e.g., group, compare).

Lesson 2: Problem Solution

Materials Needed

Checklists	Word Problem–Solving Steps (FOPS) poster
	Change Problem–Solving Checklist (laminated copies for students)
Diagram	Change Problem diagram poster
Reference Guide	Lesson 2: Change Reference Guide 1
Overhead Modeling	Lesson 2: Change Problems 1, 2, and 3
Student Pages	Lesson 2: Change Problems 1, 2, and 3

Teacher: Today we are going to use diagrams like the ones you learned earlier to solve change word problems. Let's review the change problem. A change problem has a beginning, a change, and an ending. The beginning, change, and ending all describe the same thing or object. [*Display Word Problem–Solving Steps* (FOPS) *poster. Point to each step on the poster and read each one.*] We will use these four steps to solve all addition and subtraction word problems. When we put the first letter of each step together, the letters make a funny word, FOPS. Let's review the four steps: F—Find the problem type; O—Organize the information using a diagram; P—Plan to solve the problem; S—Solve the problem. If you remember FOPS, it can help you recall the four steps. What are the four steps?

Students: F—Find the problem type; O—Organize the information using a diagram; P—Plan to solve the problem; S—Solve the problem.

Change Problem 1

Teacher: (*Display Overhead Modeling page for Change Problem 1. See Change Reference Guide 1 to set up the problem. Pass out student copies of Change Problem 1 and laminated copies of Change Problem–Solving Checklist. Point to Change Problem–Solving Checklist.*) We will use this checklist that has the same four steps (FOPS) to help us solve change problems.

We are ready for Step 1: Find the problem type. (*Point to first check box on Change Problem–Solving Checklist.*) To find the problem type, I will read the problem and retell it in my own words. Follow along as I read Problem 1. (*Read Change Problem 1 aloud.*)

"Tammy likes to paint pictures of flowers. She has painted 12 pictures so far. If she paints 4 more pictures, how many will she have?"

Now I'll retell the problem in my own words to help me understand it. When I retell, I will ask myself, What do I know in this problem, and what am I asked to find out? (*Retell the problem.*)

I know that Tammy began by painting 12 pictures. I also know that she might paint 4 more pictures. I don't know how many pictures she will have at the end if she paints more pictures. I need to solve for this ending amount.

I read the problem and told it in my own words. I will check off the first box under Step 1 on the checklist. (*Point to second check box under Step 1.*) Now I will ask myself if the problem is a change problem. Why do you think this is a change problem? What does the problem describe?

Students: It has a beginning, a change, and an ending. They all describe pictures.

Teacher: Let's check off the second box under Step 1. Now I am ready for Step 2: Organize the information in the problem using the change diagram. (*Display Change Problem diagram poster.*)

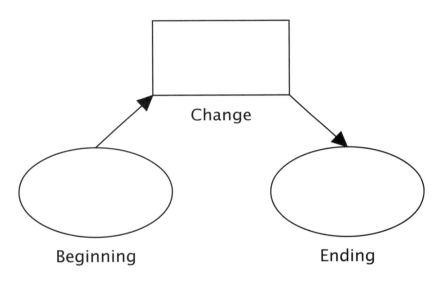

Change Problem

(*Point to first box under Step 2 in Change Problem–Solving Checklist.*) To organize the information in a change problem, I need to first underline the label that talks about the beginning, change, and ending. What is the label that describes the beginning, change, and ending in this problem?

Students: Pictures.

Teacher: Good, let's underline <u>pictures</u> and write the label in the diagram for the beginning, change, and ending. (*Pause for students to complete.*) Let's check off the first box under Step 2 on the checklist.

(*Point to second box under Step 2 on Change Problem–Solving Checklist.*) Now let's read the story to underline and circle information about the beginning, change, and ending. The first sentence says, "Tammy likes to paint pictures of flowers." Does this sentence tell about the beginning, change, or ending?

Students: No.

Teacher: Let's cross it out, because we don't need this information to solve the problem. The next sentence says, "She has painted (*underline*) 12 pictures so far." Does this sentence tell us about the beginning, change, or ending? How do you know?

Students: Beginning, because the words has painted tell us about the number of pictures Tammy started with.

Teacher: Underline has painted. (*Pause for students to complete.*) Do you know the beginning amount from this sentence?

Students: Yes, 12.

Teacher: Circle "12" and write it in for the beginning amount. The next sentence says, "If she paints 4 more pictures (*underline*), how many will she have?" This sentence has two parts to it. It describes both the change and the ending. What is the change in this problem? How do you know?

Students: The words "paints more" talk about the change action.

Teacher: Underline paints more. Circle "4," and write in "+ 4" for the change amount in the diagram. (*Pause for students to complete.*) This sentence also asks the question, "How many will she have?" (*Point to third box under Step 2 on checklist.*) The words "will have (now)" talk about the ending. Underline will have and write a "?" for it in the diagram, because we don't know this ending amount.

We underlined the important information, circled the numbers, and wrote them in the diagram. We also wrote a "?" for the ending amount we need to solve. Let's check off the second and third boxes under Step 2 on the checklist. Now, let's look at the diagram and read what it says: Tammy began with 12 pictures. If she paints 4 more pictures, we have to find out how many pictures she will then have. What must you solve for in this problem? (Is it the beginning, change, or ending amount?)

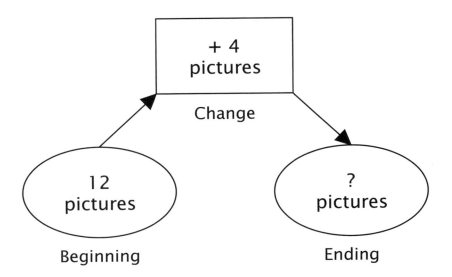

Students: The ending amount.

Teacher: Now for Step 3: Plan to solve the problem. (*Point to first box under Step 3 on checklist.*) To plan how to solve the problem, we first decide whether to add or subtract. To do that, I must ask if the total (i.e., whole) is given. Because the change (+ 4) indicates an increase to the beginning amount, the ending amount is the greater quantity (i.e., total or whole). I will write a "T" for the ending to remind me that this is the total. From this diagram, I don't know the total (i.e., the ending amount). If the total or whole is not given, we add; if the total or whole is given, we subtract. (*Point to diagram.*) The total in this problem is not given. So do we add or subtract?

Students: Add.

Teacher: That's right. We add, because we need to solve for the total (parts make up the whole or total). Let's check off the first box under Step 3 on the checklist. (*Point to second box under Step 3 on checklist.*)

Next we write the math sentence. Remember, we have to add to find the total or whole. Also, remember to line up the numbers correctly. So the math sentence is 12 + 4 = ?. Let's write the math sentence and check off the second box under Step 3 of the checklist. (*Pause for students to complete.*)

Now we are ready for Step 4: Solve the problem. (*Point to first box under Step 4.*) What does 12 + 4 =? (*Pause for students to solve the problem.*)

Students: 16.

Teacher: Check off the first box under Step 4 on the checklist. (*Point to second box under Step 4.*) Now write "16" for the "?" in the diagram and write the complete answer on the answer line. The complete answer is the number and the label. What is the complete answer to this change problem?

Students: 16 pictures.

Teacher: Good. I'll write 16 pictures on the answer line. (*Pause for students to write the answer.*) Let's check off the second box under Step 4 on the checklist. (*Point to third box under Step 4 on checklist.*)

We are now ready to check the answer. Does "16" seem right? Yes, because Tammy should end up with more pictures (16) than she started with (12). We can also check by subtracting: 16 − 12 = 4; 16 − 4 = 12. (*Check off third box under Step 4.*)

(*Model writing the explanation for how the problem was solved here; see Change Reference Guide 1.*) Let's review this change problem: (1) There is a beginning, a change, and an ending (*point to Change Problem diagram*), and (2) we began with pictures and ended with pictures. The change also involved pictures (*point to diagram*). What's this problem called? Why?

Students: Change, because it has a beginning, a change, and an ending amount. They all describe pictures.

Change Problem 2

Teacher: (*Display Overhead Modeling page for Change Problem 2. Have students look at student page copies of Change Problem 2.*)

Touch Problem 2. (*Point to Change Problem–Solving Checklist.*) What's the first step?

Students: Find the problem type.

Teacher: Right. (*Point to first check box on checklist.*) To find the problem type, I will read the problem and retell it in my own words. (*Read Change Problem 2 aloud.*)

"Brian bought hot dogs for a picnic. He gave the cashier $20. He got $6 back. How much did the hot dogs cost?"

I read the problem. What must I do next?

Students: Retell the problem using own words.

Teacher: Yes, I will retell the problem in my own words to help me understand it. When I retell, I will ask myself, What do I know in this problem, and what am I asked to find out? (*Retell aloud.*)

I know that Brian began with $20. This is the beginning amount. Then he bought some hot dogs. We don't know how much the hot dogs cost, which is the change amount. He got back $6 from the cashier. This is the amount he now has or the ending amount.

I read the problem and told it in my own words. Check off the first box under Step 1 on the checklist.

(*Point to second check box under Step 1.*) Now I will ask myself if it is a change problem. Why do you think this is a change problem? What do they all describe?

Students: It has a beginning, a change, and an ending. They all describe dollars.

Teacher: Check off the second box under Step 1. Now I am ready for Step 2: Organize the information in the problem using the change diagram. (*Display Change Problem diagram poster. Point to first check box under Step 2 in Change Problem–Solving Checklist.*)

To organize the information, you need to first underline the label that talks about the beginning, change, and ending. What is the label that describes the beginning, change, and ending in this problem?

Students: Dollars.

Teacher: Good, we will underline $ and write it in the diagram for the beginning, change, and ending. (*Pause for students to write.*) Let's check off the first box under Step 2 on the checklist.

(*Point to second check box under Step 2 on checklist.*) Now let's read the problem and underline the information about the beginning, change, and ending. The first sentence says, "Brian bought hot dogs for a picnic." I'll underline the word bought because it describes the change action in the problem. (*Pause for students to complete.*) However, this sentence does not give me the change amount for the hot dogs, so I will read the next sentence.

The second sentence says, "He gave the cashier $20." The sentence tells us that Brian had $20 (beginning amount) before he bought (i.e., a change action) the hot dogs. (*Emphasize that we have to infer the beginning amount in this sentence, because it is not clearly stated in the problem.*) Circle "$20," and write it for the beginning amount in the diagram. (*Pause for students to complete.*)

The next sentence says, "He got $6 back." This is the amount he now has. (*Emphasize again that we have to infer the ending amount in this sentence, because it is not clearly stated in the problem.*) This sentence tells me about an ending amount of $6. I will circle "$6" and write it in the diagram for the ending amount.

The last sentence is the question. (*Point to third check box under Step 2 on checklist.*) It asks, "How much did the hot dogs cost?" This sentence refers to the change amount (i.e., he had $20 and then he bought some hot dogs for the picnic) that we need to solve. I don't know this amount, so I will write a "– ?," because when you pay for something, you have less than what you started with.

We underlined the important information, circled the numbers, and wrote them in the diagram. We also wrote a "– ?" for the change amount we need to solve. Let's check off the second and third boxes under Step 2 on the checklist.

Now let's look at the diagram and read what it says. Brian began with $20, bought some hot dogs, and he now has $6. We need to find out how much he paid for the hot dogs. What must you solve for in this problem? (Is it the beginning, change, or ending amount?)

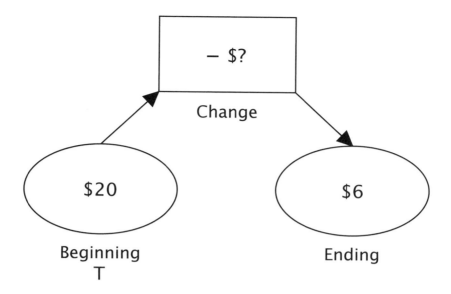

Students: The change amount.

Teacher: Now for Step 3: Plan to solve the problem. (*Point to first box under Step 3 on checklist.*) To plan how to solve the problem, we first decide whether to add or subtract. To do that, I must ask if the total (i.e., whole) is given. Because the change (− ?) indicates a decrease to the beginning amount, the beginning amount ($20) is the greater quantity (i.e., total). I will write a "T" for the beginning to remind me that this is the total. From this diagram, I know the total (i.e., the beginning amount). If the total or whole is not given, we add; if the total or whole is given, we subtract. The total in this problem is given. (*Point to diagram.*) So do we add or subtract?

Students: Subtract.

Teacher: That's right. We know the total (beginning amount) and one of the parts (ending amount), so we subtract to solve for the unknown part (i.e., change). Let's check off the first box under Step 3 on the checklist.

(*Point to second check box under Step 3 on checklist.*) Next we write the math sentence. Remember we have to subtract, so we start with the total or whole and subtract the other number (part). Also, remember to line up the numbers correctly. So the math sentence is $20 − $6 = ?. (*Remind students to write in the dollar sign before the numbers.*) Let's write the math sentence and check off the second box under Step 3 of the checklist. (*Pause for students to complete.*)

Now we are ready for Step 4: Solve the problem. (*Point to first box under Step 4 on checklist.*) What does $20 − $6 =? (*Pause for students to solve problem.*)

Students: $14.

Teacher: Check off the first box under Step 4 on the checklist. (*Point to second check box under Step 4 on checklist.*) Now write $14 for the "– ?" in the diagram and write the complete answer on the answer line. The complete answer is the number and the label. What is the complete answer to this change problem?

Students: $14.

Teacher: Good. I'll write $14 on the answer line. (*Pause for students to write answer.*) Let's check off the second box under Step 4 on the checklist. (*Point to third check box under Step 4 on checklist.*) We are now ready to check the answer. Does $14 seem right?

Students: Yes, because Brian ended with less ($6) than he started with ($20).

Teacher: We can also check by adding: $14 + $6 = $20. (*Check off the third check box under Step 4. Next guide students to write the explanation for solving the problem on their worksheet; see Change Reference Guide 1.*)
   Let's review this change problem: (1) There is a beginning, a change, and an ending (*point to the diagram*), and (2) we began with dollars and ended with dollars. The change also describes dollars. (*Point to diagram.*) What's this problem called? Why?

Students: Change, because it has a beginning, a change, and an ending amount. They all begin with dollars and end with dollars. The change also describes dollars.

Change Problem 3

Teacher: (*Use this script as a guideline for solving Change Problems 3 and 4, and facilitate problem solving by having frequent student–teacher exchanges. Display Overhead Modeling page for Change Problem 3. Have students look at student pages of Change Problem 3.*)
   Touch Problem 3. What's the first step?

Students: Find the problem type.

Teacher: Right. (*Point to first check box on Change Problem–Solving Checklist.*) To find the problem type, what must you do?

Students: Read the problem and retell it in your own words.

Teacher: Good. Read the problem aloud.

Students: "The zoo tour bus is pulling into the Fairview stop. 14 people get on the bus. Now there are 35 people on the bus. How many people were on the bus before the Fairview stop?"

Teacher: You read the problem. What must you do next?

Students: Retell it using own words.

Teacher: Yes. Now retell the problem in your own words to help you understand it. (*Call on students to retell the problem. When they retell the problem, remind students to tell what they know in the problem and what they are asked to find out.*)

Students: The zoo tour bus began with some people already on it. Then it picked up 14 more people at the Fairview stop. Now the bus has 35 people in it. I need to solve for the beginning amount (i.e., the number of people on the bus when it began the tour).

Teacher: (*Check off first box under Step 1 on checklist. Point to second check box under Step 1.*) What kind of a problem is this? How do you know?

Students: It is a change problem, because the words "14 people get on" tell about a change, and the words "now there are 35 people" tell about the ending amount. We have to find the beginning amount or the number of people already on the bus when it pulled into the Fairview stop.

Teacher: Check off the second box under Step 1 on the checklist. Now you are ready for Step 2: Organize the information in the problem using the change diagram. (*Point to first box under Step 2 of checklist.*) To organize the information in a change problem, what must you do first?

Students: Underline the label for the beginning, change, and ending amounts.

Teacher: What is the label in this problem?

Students: People (on the zoo tour bus).

Teacher: Good. Underline <u>people</u> in the problem, and write "people" for the beginning, change, and ending in the change diagram. (*Pause for students to write.*) Let's check off the first box under Step 2 on the checklist. (*Point to second box under Step 2 on checklist.*) What do you do next?

Students: Read the problem and underline the information given about the beginning, change, and ending.

Teacher: The first sentence says, "The zoo tour bus is pulling into the Fairview stop." Does this sentence tell about the beginning, change, or ending?

Students: No.

Teacher: Let's cross it out, because we don't need this information to solve the problem. The next sentence says, "14 people get on the bus." What does this sentence describe? How do you know?

Students: A change amount, because the words "get on" describe a change action.

Teacher: Is the change an increase or a decrease in quantity? How do you know?

Students: An increase, because when people get on the bus, there are more of them.

Teacher: Circle "14" and write in "+ 14" for the change in the diagram. (*Pause for students to complete.*) The next sentence says, "Now there are 35 people on the bus." What does this describe? How do you know?

Students: Ending amount, because that's the number of people now on the bus.

Teacher: The last sentence is the question. (*Point to third check box under Step 2 on checklist.*) It asks, "How many people <u>were on the bus</u> (*underline*) before the Fairview stop?" What does this question ask us to solve? How do you know?

Students: The beginning amount, because the words "were on the bus" refer to the number of people who were already on the bus (before the Fairview stop).

Teacher: We don't know this amount, so write a "?" for it. We underlined the important information, circled the numbers, and wrote them in the diagram. We also wrote a "?" for the beginning amount we need to solve. Let's check off the second and third boxes under Step 2 on the checklist.

Now look at the diagram and read what it says. There were some people on the bus at the beginning. Then 14 people got on the bus. Now there are 35 people. What must you solve for in this problem? (Is it the beginning, change, or ending amount?)

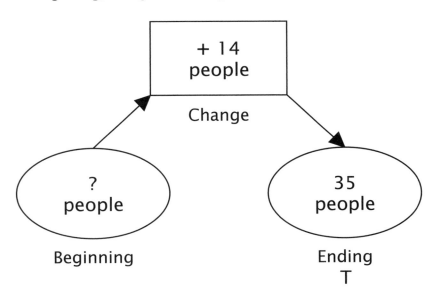

Students: Beginning amount.

Teacher: Good. Now for Step 3: Plan to solve the problem. To plan how to solve the problem, what must you do? (*Point to first check box under Step 3 on checklist.*)

Students: Decide whether to add or subtract.

Teacher: To figure out whether to add or subtract, what should you ask yourself?

Students: If the total or whole is given or not given.

Teacher: Right! In this change problem, is the total the beginning or ending amount? How do you know?

Students: Ending amount, because the change (+ 14) indicates an increase to the beginning amount.

Teacher: Write "T" for the ending amount. (*Pause for students to write.*) From this diagram, do you know the total (ending amount)?

Students: Yes.

Teacher: If the total is given, do you add or subtract?

Students: Subtract.

Teacher: Good. Check off the first box under Step 3 on the checklist. What do you do next? (*Point to second check box under Step 3 on checklist.*)

Students: Write the math sentence.

Teacher: Remember, you have to subtract, so start with the total or whole and subtract the other number. Also, remember to line up the numbers correctly. What is the math sentence?

Students: 35 − 14 = ?.

Teacher: Write it and check off the second box under Step 3 on the checklist. (*Pause for students to complete.*) Now you are ready for Step 4: Solve the problem. What do you do first?

Students: Solve for 35 − 14.

Teacher: What does 35 − 14 =? (*Pause for students to solve problem.*)

Students: 21.

Teacher: Check off the first box under Step 4 on the checklist. What do you do next? (*Point to second check box under Step 4.*)

Students: Write the complete answer.

Teacher: The complete answer is the number and the label. What is the complete answer to this change problem?

Students: 21 people.

Teacher: Good. Write 21 for the "?" in the diagram, and write 21 people on the answer line. (*Pause for students to write answer.*) Let's check off the second box under Step 4 on the checklist. (*Point to third check box under Step 4 on checklist.*) What do you do next?

Students: Check the answer.

Teacher: Does 21 seem right?

Students: Yes, because the tour bus ended up with more people (35) than it started with (21).

Teacher: We can also check by adding: 21 + 14 = 35. Check off the third box under Step 4. Let's review this change problem: (1) There is a beginning, a change, and an ending (*point to diagram*), and (2) we began with people (on the bus) and ended with people. The change also involved people (*point to diagram*). What is this problem called? Why?

Students: Change, because it has a beginning, a change, and an ending. They all describe people.

Teacher: Great job! Tomorrow we will practice more change problems.

Lesson 3: Problem Solution

Materials Needed

Answer Sheet to Paired-Learning Tasks	Lesson 3: Change Answer Sheet 1
Checklists	Word Problem–Solving Steps (FOPS) poster
	Change Problem–Solving Checklist (laminated copies for students)
Diagram	Change Problem diagram poster
Overhead Modeling	Lesson 3: Change Problem 4
Student Pages	Lesson 3: Change Problem 4
	Lesson 3: Change Worksheet 1

Change Problem 4

Teacher: [*Display Word Problem–Solving Steps (FOPS) poster. Ask students to read each step on the poster, which they will use to solve addition and subtraction word problems. Display Overhead Modeling page for Change Problem 4. Pass out student pages for Change Problem 4.*]
Touch Problem 4. (*Point to Change Problem–Solving Checklist.*) Remember, we will use this checklist that has the same four steps (FOPS) to help us solve change problems. What's the first step?

Students: Find the problem type.

Teacher: Right. (*Point to first check box on checklist.*) To find the problem type, what must you do?

Students: Read the problem and retell it in your own words.

Teacher: Good. Read the problem aloud.

Students: "It takes 6 hours to drive to Boston. Your mother tells you that there are 2 hours of driving left before you get to Boston. How many hours have you been in the car?"

Teacher: You read the problem. What must you do next?

Students: Retell the problem using own words.

Teacher: Yes. Now retell the problem in your own words to help you understand it. (*Call on students to read and retell the problem. When they retell the problem, remind students to tell what they know in the problem and what they are asked to find out.*)

Students: I know that when we started, it was 6 hours to Boston. This is the beginning amount. We have been driving for some time, and now it is 2 hours to Boston. This is the ending amount. I don't know how many hours we have driven or been in the car. This is the change amount that I must solve in this problem.

Teacher: Good. Check off the first box under Step 1 on the checklist. (*Point to second check box under Step 1.*) What kind of a problem is this? How do you know?

Students: It is a change problem, because it has a beginning, a change, and an ending. They all talk about hours.

Teacher: Check off the second box under Step 1 on the checklist. Now you are ready for Step 2: Organize the information in the problem using the change diagram. To organize the information, what must you first underline and write in the change diagram?

Students: The label for the beginning, change, and ending.

Teacher: What is the label that describes the beginning, change, and ending?

Students: Hours.

Teacher: Good. Underline hours in the problem, and write it in for the beginning, change, and ending in the diagram. Check off the first box under Step 2 on the checklist. (*Point to second box under Step 2 on checklist.*)
Now read the problem and underline the information given about the beginning, change, and ending. The first sentence says, "It takes 6 hours to drive to Boston." Does this sentence tell about the beginning, change, or ending? How do you know?

Students: The beginning, because it tells us about the total number of hours of driving time to Boston when we began the car trip.

Teacher: Circle "6," and write it in for the beginning amount in the diagram. (*Pause for students to complete.*) The next sentence says, "Your mother tells you that there are 2 hours of driving left before you get to Boston." Does this sentence tell about the change or ending amount? How do you know?

Students: The ending amount, because it tells us that it is now 2 hours to Boston.

Teacher: Circle "2" and write it in for the ending amount in the diagram. (*Pause for students to complete.*) The last sentence is a question. (*Point to third box under Step 2 on checklist.*) It asks, "How many hours have you been (*underline*) in the car?" What does this question ask us to solve? How do you know?

Students: Change amount, because the words "have been in the car" tell about the number of hours driven after we started our trip.

Teacher: We don't know this amount. If we have been driving for a few hours, does the change involve an increase or decrease to the beginning amount of 6 hours?

Students: Decrease.

Teacher: Right! Write a "– ?" for the change amount. (*Pause for students to complete.*) We underlined the important information, circled numbers, and wrote the numbers in the diagram. We also wrote a "?" for the change amount we need to solve. Let's check off the second and third boxes under Step 2 on the checklist. Now read what the diagram says (e.g., When we started, it was 6 hours to Boston. Then we drove for a few hours. Now it is 2 hours to Boston.). What must you solve for in this problem? (Is it the beginning, change, or ending amount?)

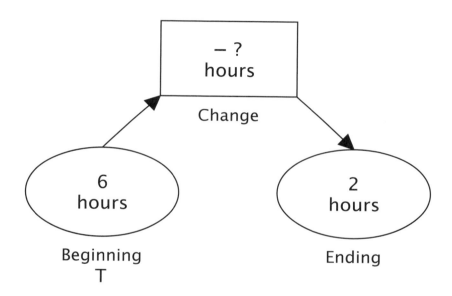

Students: Change amount.

Teacher: Good. Now for Step 3: Plan to solve the problem. To plan how to solve the problem, what must you do? (*Point to first box under Step 3 on checklist.*)

Students: Decide whether to add or subtract.

Teacher: To figure out whether to add or subtract to solve this problem, what should you ask yourself?

Students: If the total or whole is given or not given.

Teacher: Right! In this problem (*point to diagram*), is the total the beginning or ending amount? How do you know?

Students: Beginning amount, because the change (− ? hours) indicates a decrease to the beginning amount.

Teacher: Write "T" for the beginning amount. (*Pause for students to write.*) From this diagram, do you know the total (beginning amount)?

Students: Yes.

Teacher: If the total is given, do you add or subtract?

Students: Subtract.

Teacher: Good. Check off the first box under Step 3 on the checklist. What do you do next? (*Point to second box under Step 3 on checklist.*)

Students: Write the math sentence.

Teacher: Remember, you have to subtract, so start with the total and subtract the other number. Also, remember to line up the numbers correctly. What is the math sentence?

Students: 6 − 2 = ?.

Teacher: Write it and check off the second box under Step 3 on the checklist. (*Pause for students to complete.*) Now you are ready for Step 4: Solve the problem. What do you do first?

Students: Solve for 6 − 2.

Teacher: What does 6 − 2 =? (*Pause for students to solve problem.*)

Students: 4.

Teacher: Check off the first box under Step 4 on the checklist. What do you do next? (*Point to second box under Step 4.*)

Students: Write the complete answer.

Teacher: The complete answer is the number and the label. What is the complete answer to this change problem?

Students: 4 hours.

Teacher: Good. Write "4" for the "?" in the diagram, and write "4 hours" on the answer line. (*Pause for students to write the answer.*) Let's check off the second box under Step 4 on the checklist. (*Point to third box under Step 4 on checklist.*) What do you do next?

Students: Check the answer.

Teacher: Does 4 seem right?

Students: Yes, because the drive to Boston is now (2 hours) less than when we started (6 hours).

Teacher: We can also check by adding: 4 + 2 = 6. Check off the third box under Step 4. Let's review this change problem: (1) There is a beginning, a change, and an ending (*point to diagram*), and (2) we began with hours and ended with hours. The change also involved hours (*point to diagram*). What is this problem called? Why?

Students: Change, because it has a beginning, a change, and an ending. They all describe hours.

Teacher: (*Pass out Change Worksheet 1.*) Now I want you to do Problem 1 on this worksheet with your partner. (*Ask students to* think, plan, *and* share *with partners to solve Change Worksheet Problem 1; see Guide to Paired Learning section in the Introduction.*)

Change Worksheet 1, Problem 1: "Bobby prepared a lot of meals for his long hike. He ate 42 meals as he hiked through the mountains. There were 15 meals left at the end of his hike. How many meals did he prepare for his hike?"

(*Monitor students as they work. Have students check their answers using Change Answer Sheet 1. Make sure the diagram is labeled correctly, the total is written, the math sentence is written and worked out correctly, the written explanation is complete, and the complete answer is written on the answer line.*)

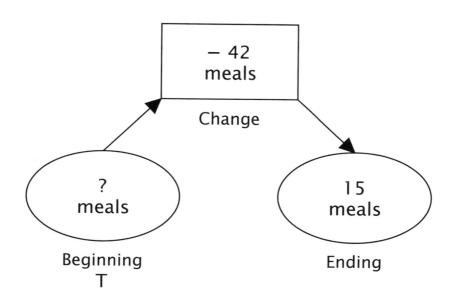

Answer: 57 meals

Teacher: Now I want you to do the next two problems on your own. Remember to use the four steps to solve these problems.

Change Worksheet 1, Problem 2: "There are 25 penguins on the ice. 6 of them jump into the water. How many penguins are still on the ice?"

(Monitor students as they work. After some time, go over the answer. Make sure the diagram is labeled correctly, the total is written, the math sentence is written and worked out correctly, the written explanation is complete, and the complete answer is written on the answer line; see below.)

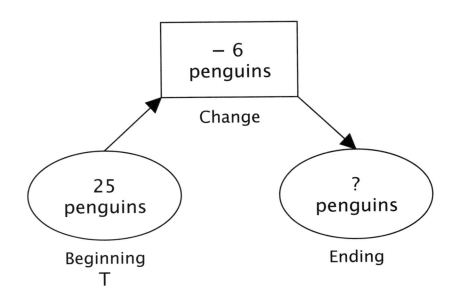

Answer: 19 penguins

Change Worksheet 1, Problem 3: "José and his father have gathered 10 pounds of wool from a sheep. So far, some of the wool has been used to make a sweater. Now there are 5 pounds of wool left. How many pounds of wool have been used?"

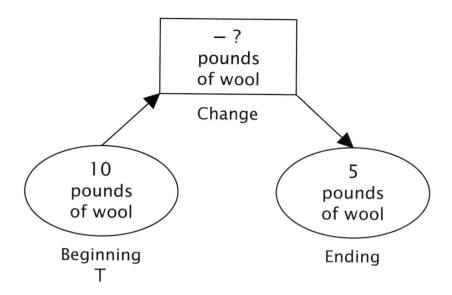

Answer: 5 pounds of wool

Teacher: Great job working hard! Tomorrow we will practice more change problems.

Lesson 4: Problem Solution

Materials Needed

Answer Sheet for Paired-Learning Tasks	Lesson 4: Change Answer Sheet 2
Checklists	Word Problem–Solving Steps (FOPS) poster
	Change Problem–Solving Checklist (laminated copies for students)
Diagram	Change Problem diagram poster
Overhead Modeling	Lesson 4: Change Worksheet 2, Problem 1
Student Pages	Lesson 4: Change Worksheet 2

Teacher: (*Pass out Change Worksheet 2. Display Overhead Modeling page of Change Worksheet 2, Problem 1.*)

The information in this problem is given in a table. We can still use our four steps to solve this problem. (*Use guided practice to have students complete the problem.*)

"Use the table below to solve Change Worksheet 2, Problem 1."

Supplies

Item	Price
Glitter	$3.50
Glue	$1.70
Hole punch	$4.00

Change Worksheet 2, Problem 1: "Gina paid for glue with a $5 bill. How much did she get back?"

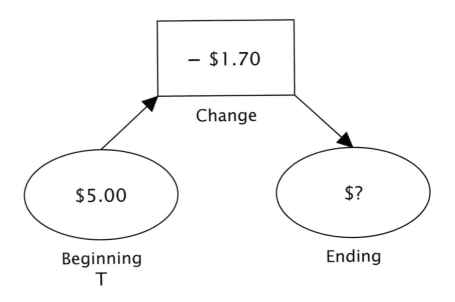

Answer: $3.30

Teacher: Now I want you to do the next problem with your partner. (*Ask students to* think, plan, *and* share *with partners to solve the problem; see Guide to Paired Learning in Introduction.*)

Change Worksheet 2, Problem 2: "A squirrel made a pile of nuts. It carried away 55 nuts to its nest. Now there are 38 nuts in the pile. How many nuts were in the pile at the beginning?"

(*Use the four steps to solve a change problem. Monitor students as they work. Have students check their answers using Change Answer Sheet 2. Make sure the diagram is labeled correctly, the math sentence is written and worked out correctly, the written explanation is complete, and the complete answer is written on the answer line.*)

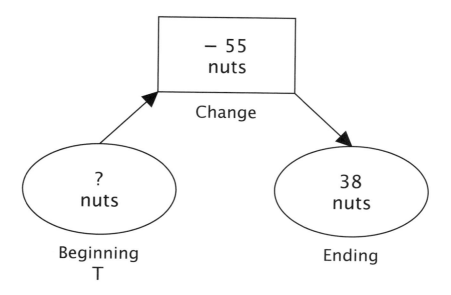

Answer: 93 nuts

Teacher: Now I want you to do the next four problems on your own. Remember to use the four steps to solve the problems on this worksheet.

Change Worksheet 2, Problem 3: "Tina bought some eggs. She used 8 eggs for breakfast. She now has 16 eggs. How many eggs did she begin with?"

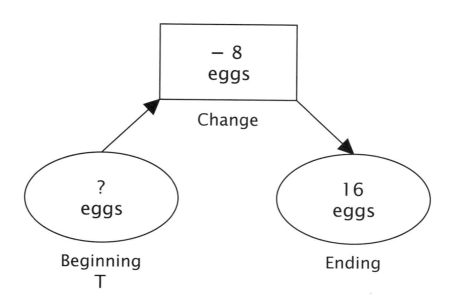

Answer: 24 eggs

Change Worksheet 2, Problem 4: "16 ducks are swimming in the water. 9 more ducks jumped into the water from the shore. How many ducks are in the water now?"

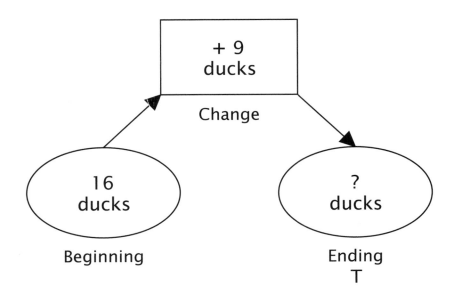

Answer: 25 ducks

Change Worksheet 2, Problem 5: "You have some trading cards, and your friend gives you 6 more. Now you have 15 trading cards. How many trading cards did you have at the beginning?"

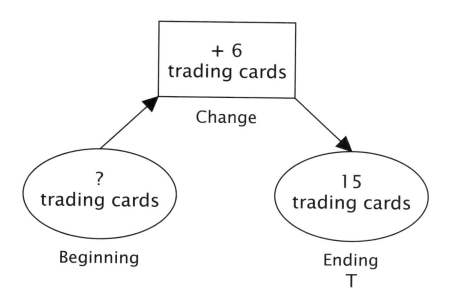

Answer: 9 trading cards

"Use the table below to solve Change Worksheet 2, Problem 6."

Bulbs by the Bag

Item	Price
Tulips	$4.50
Daffodils	$2.50
Irises	$4.00

Change Worksheet 2, Problem 6: "Lynn pays for a bag of daffodils with $10. How much does she now have?"

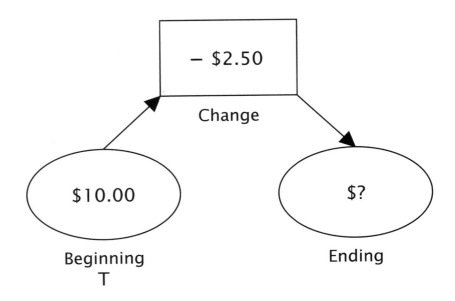

Answer: $7.50

Teacher: (*Monitor students as they work. After 15 to 20 minutes, go over the answers. Make sure diagrams are labeled correctly, totals are written, math sentences are written and worked out correctly, and complete answers are written on the answer lines.*)

Great job working hard. Tomorrow we will solve some more change problems and then get ready to learn the next problem type, a group problem.

Lesson 5: Problem Solution

Materials Needed

Checklists	Word Problem–Solving Steps (FOPS) poster
	Change Problem–Solving Checklist (laminated copies for students)
Overhead Modeling	Lesson 5: Change Worksheet 3, Problem 1
Reference Guide	Lesson 5: Change Reference Guide 2
Student Pages	Lesson 5: Change Worksheet 3

Teacher: (*Pass out Change Worksheet 3. Display Overhead Modeling page for Change Worksheet 3, Problem 1.*)

You learned to solve change word problems using diagrams. Now we will solve the problems on this worksheet using your own diagrams. This worksheet does not have diagrams. Remember to use the four steps (FOPS) to solve problems on the worksheet. (*Note: Discuss how students can generate a diagram that is more efficient than the one they used and have them practice solving the problems using the diagram they generate. Also, encourage them to use the Change Problem–Solving Checklist only as needed.*)

(*Use guided practice to complete Change Worksheet 3, Problems 1 and 2, using own diagrams; see below. Model how to read information in a pictograph for Problem 2, using Change Reference Guide 2.*)

Change Worksheet 3, Problem 1: "Yesterday Melanie pasted some pictures in her scrapbook. Today she put in 9 more pictures. Now there are 25 pictures in the scrapbook. How many pictures did she have in her scrapbook yesterday?"

$$
\begin{array}{ccc}
 & \underline{\begin{array}{c} + 9 \\ \text{pictures} \end{array}} & \\
 & C & \\
\underline{\begin{array}{c} ? \\ \text{pictures} \end{array}} & & \underline{\begin{array}{c} 25 \\ \text{pictures} \end{array}} \\
B & & E \\
 & & T
\end{array}
$$

Answer: 16 pictures

Addition and Subtraction

"Use data from the pictograph to solve Change Worksheet 3, Problem 2."

Books Read in Read-a-thon by Mrs. Blake's Class

September	📕 📕 📕 📕 📗
October	📕 📕 📗
November	📕 📕 📕 📕
December	📕 📕 📕 📕 📕 📕
January	📕 📕 📕 📕 📕 📕 📕 📕

📕 = 10 books

📗 = 5 books

Change Worksheet 3, Problem 2: "The third graders in Mrs. Blake's class read books for a Read-a-thon. How many books did they read by the end of October?" (*Discuss the pictograph, using Change Reference Guide 2.*)

$$\begin{array}{c} + 25 \\ \text{books} \\ \hline C \end{array}$$

$$\begin{array}{c} 45 \\ \text{books} \\ \hline B \end{array} \qquad \begin{array}{c} ? \\ \text{books} \\ \hline E \\ T \end{array}$$

Answer: 70 books

Teacher: Now I want you to do the next three problems on your own. (*Have students write the explanation for at least one of the three problems.*) Remember to use the four steps to solve Problems 3 through 5 on Worksheet 3.

(*Monitor students as they work. After some time, go over the answers. Make sure diagrams are labeled correctly, totals are written, math sentences are written and worked out correctly, and complete answers are written on the answer lines; see below.*)

Change Worksheet 3, Problem 3: "You prepared a pile of blocks for making a model. Then your friend borrowed 41 blocks from you. Now you have 55 blocks left. How many blocks did you prepare at the beginning for the model?"

$$\underset{C}{\underline{\begin{array}{c}-\ 41\\ \text{blocks}\end{array}}}$$

$$\underset{\begin{array}{c}B\\T\end{array}}{\underline{\begin{array}{c}?\\ \text{blocks}\end{array}}} \qquad\qquad \underset{E}{\underline{\begin{array}{c}55\\ \text{blocks}\end{array}}}$$

Answer: 96 blocks

Change Worksheet 3, Problem 4: "You had 32 French fries. You got 15 more French fries from your sister. How many French fries do you have now?"

$$\underset{C}{\underline{\begin{array}{c}+\ 15\\ \text{French fries}\end{array}}}$$

$$\underset{B}{\underline{\begin{array}{c}32\\ \text{French fries}\end{array}}} \qquad\qquad \underset{\begin{array}{c}E\\T\end{array}}{\underline{\begin{array}{c}?\\ \text{French fries}\end{array}}}$$

Answer: 47 French fries

"Use the table below to solve Change Worksheet 3, Problem 5."

Third-Grade Students at County Elementary School

Ms. Griffin's classroom	20
Mrs. Smith's classroom	19
Mr. Chard's classroom	18
Ms. Howard's classroom	16

Change Worksheet 3, Problem 5: "Suppose 5 new students will be entering Ms. Howard's class next year, and none of the current students leave. How many students will there be in Ms. Howard's class?"

$$\frac{+\ 5\ \text{students}}{C}$$

$$\frac{16\ \text{students}}{B} \qquad \frac{?\ \text{students}}{\begin{array}{c}E\\T\end{array}}$$

Answer: 21 students

Teacher: You have learned to solve *change* word problems using your own diagrams. Next you will learn to map information from *group* word problems and solve them.

Unit 2

Group Problems

Lesson 6: Problem Schema

Materials Needed

Checklist	Group Story Checklist (laminated copies for students)
Diagram	Group Problem diagram poster
Overhead Modeling	Lesson 6: Group Stories 1, 2, and 3
Student Pages	Lesson 6: Group Stories 1, 2, and 3
	Lesson 6: Group Schema Worksheet 1

Teacher: Today we will learn to identify and organize another type of addition and subtraction problems called "group," so that we can later solve them. Group problems will help you to understand the part–part–whole relationship. That is, the large group or whole is equal to the sum of the smaller groups or parts. In a group problem, the large group or whole tells about all the things in the problem. I'll say some group names. You name the large group and tell me the small groups that are part of the large group. Listen: dogs, pets, hamsters. What is the large group or total?

Students: Pets.

Teacher: What are the small groups or parts?

Students: Dogs and hamsters.

Teacher: (*Repeat with these groups: girls, boys, children; brown and not brown eggs [or all eggs], not brown eggs, brown eggs; pencils, yellow pencils, green pencils, red pencils, black pencils; cars, trucks, vehicles; Ed's baseball cards, Joe's baseball cards, Ed's and Joe's baseball cards [all baseball cards]; chocolate milk, white milk, chocolate and white milk; tickets sold on Monday, tickets sold on Friday, tickets sold on Monday and Friday [all tickets sold].*)

Group Story 1

Teacher: (*Display Group Story Checklist.*) Here are two steps we will use to organize information in a group story.

(*Point to each step on the Group Story Checklist and read each one.*) Let's use these two steps to do an example. Look at this story. (*Display Overhead Modeling page for Group Story 1. Pass out student copies of Group Stories 1, 2, and 3 and Group Story Checklist.*)

(*Point to the first check box on the Group Story Checklist.*) Now we are ready for Step 1: Find the problem type. To find the problem type, I will read the story and retell it in my own words. (*Read story aloud.*)

"Ann has 3 apples and 2 oranges. She has 5 pieces of fruit."

Now I'll retell the problem in my own words to help me understand it. When I retell, I will ask myself, What do I know in this story? (*Retell the story.*)

Ann has 5 pieces of fruit altogether; 3 are apples and 2 are oranges. I read the story and told it in my own words. Let's check off the first box under Step 1 on the checklist.

(*Point to second check box under Step 1.*) Now I will ask myself if the story is a group problem. This is a group problem, because it has two small groups (e.g., apples and oranges) or parts that combine to make a large group (fruit) or whole. Also, the whole (5 fruits) is equal to the sum of the parts (3 + 2 = 5). (*Check off second box under Step 1.*)

Now I am ready for Step 2: Organize the information in the story using the group diagram. (*Display Group Problem diagram poster.*)

Group Problem

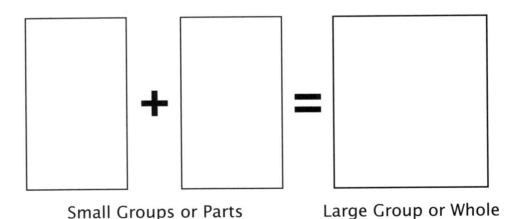

Small Groups or Parts   Large Group or Whole (Total)

(*Point to first check box under Step 2.*) To organize the information in a group problem, we first underline the words and then write the group names in the diagram. In this story, we have two small groups or parts. What are the two small groups in this story?

Students: Apples and oranges.

Teacher: Good. Underline apples and oranges in the story and write "apples" for one of the small groups and "oranges" for the other small group in the diagram. (*Pause for students to write.*) What is the large group or whole?

Students: All fruit.

Teacher: Excellent! Underline fruit in the story and write "fruit" for the large group. (*Pause for students to write.*) Let's check off the first box under Step 2 on the checklist. (*Point to second box under Step 2.*) Now let's read the story and circle the numbers given for each of the small groups and the large group. The first sentence says, "Ann has 3 apples and 2 oranges." Does this sentence tell about the large group or the small groups? How do you know?

Students: Small groups, because apples and oranges are part of fruit.

Teacher: Good. Circle "3" and "2." Write "3" for "apples" and "2" for "oranges" in the diagram for the small groups. (*Pause for students to write.*) The next sentence says, "She has 5 pieces of fruit." Does this sentence tell about the large group or the small groups? How do you know?

Students: The large group, because fruit consists of apples and oranges.

Teacher: Right, "fruit" tells about all the things (apples and oranges) in this story. Circle "5" and write it for "fruit" in the diagram. (*Pause for students to write.*) We underlined the groups, circled numbers, and wrote group names and numbers in the group diagram. Check off the second box under Step 2 on the checklist. Now let's look at the diagram and read what it says.

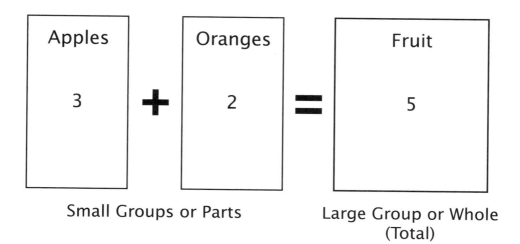

(*Point to relevant parts of the diagram as you explain.*) The story has two small groups (3 apples and 2 oranges) that combine to make a large group (5 fruits). Also, the whole (5 fruits) is equal to the sum of the parts (3 apples and 2 oranges). (*Point to the diagram.*) That is, 5 (fruits) = 3 (apples) + 2 (oranges). What is this problem called? Why?

Students: Group, because two small groups (apples and oranges) combine to make a large group (fruit). Also, the whole (5) is equal to the sum of the parts (3 + 2 = 5).

Group Story 2

Teacher: (*Display Overhead Modeling page for Group Story 2. Have students look at student pages of Group Story 2.*)

Touch Story 2. (*Point to checklist.*) What's the first step?

Students: Find the problem type.

Teacher: (*Point to first check box on group checklist.*) To find the problem type, I will read the story and retell it in my own words. (*Read story aloud.*)

"Ken and Ross are selling candy bars at a fund-raiser. Ken sold 27 candy bars. Ross sold 43 candy bars. Ken and Ross together sold 70 candy bars."

I read the story. What must I do next?

Students: Retell the story using own words.

Teacher: Yes, I will retell the story in my own words to help me understand it. When I retell, I will ask myself, What do I know in this story? (*Retell the story.*)

Together, Ken and Ross sold 70 candy bars. Ken sold 27 candy bars and Ross sold 43 candy bars.

I read the story and told it in my own words. Let's check off the first box under Step 1 on the checklist. (*Point to second check box under Step 1.*) Now I will ask myself if the story is a group problem. I know that this is a group problem. Can you tell me why this is a group problem?

Students: This is a group problem because there are two small groups (i.e., candy bars sold by Ken and candy bars sold by Ross) that combine to make the large group (all candy bars sold by Ken and Ross).

Teacher: That's right. Also, the whole (70) is equal to the sum of the parts (43 + 27 = 70). (*Check off the second box under Step 1.*) Now I am ready for Step 2: Organize the information in the story using the group diagram.

Group Problem

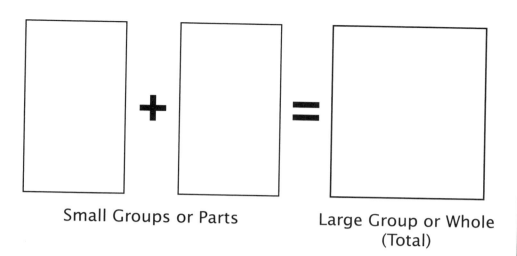

Small Groups or Parts   Large Group or Whole (Total)

(*Point to first box under Step 2.*) To organize the information in a group problem, we first underline the group names and write them in the diagram. In this story, how many small groups do we have? What are they?

Students: Two small groups. Candy bars sold by Ken and candy bars sold by Ross.

Teacher: Good, underline them in the story and write "candy bars sold by Ken" for one of the small groups and "candy bars sold by Ross" for the other small group in the diagram. (*Pause for students to write.*) What is the large group or whole?

Students: All candy bars sold by Ken and Ross.

Teacher: Excellent! Underline <u>candy bars sold by Ken and Ross</u> in the story and write it for the large group. (*Pause for students to write.*) Now check off the first box under Step 2 on the checklist.

(*Point to second box under Step 2.*) Next let's read the story and circle the numbers given for each of the small groups and the large group. The first sentence says, "Ken and Ross are selling candy bars at a fund-raiser." This does not give me any information about the groups. I will cross out this sentence. The next sentence says, "Ken sold 27 candy bars." Does this sentence tell about the large group or the small groups? How do you know?

Students: One of the small groups, because it tells us the number of candy bars that Ken sold.

Teacher: Good. Circle "27" and write it for "candy bars sold by Ken" in the diagram for one of the small groups. (*Pause for students to write.*) The next sentence says, "Ross sold 43 candy bars." Does this sentence tell about the large group or a small group? How do you know?

Students: A small group, because it tells us the number of candy bars that Ross sold.

Teacher: Good. Circle "43" and write it for "candy bars sold by Ross" in the diagram for the other small group. (*Pause for students to write.*) The last sentence tells us, "Ken and Ross together sold 70 candy bars." Does this sentence tell about the large group or the small group? How do you know?

Students: The large group, because it tells us the number of all candy bars that Ken and Ross sold together.

Teacher: Right, "candy bars that Ken and Ross sold together" tells about all the things in this story. Circle "70" in the sentence and write it for the large group in the diagram. (*Pause for students to write.*) We underlined the groups, circled the numbers, and wrote group names and numbers in the group diagram. Let's check off the second box under Step 2 on the checklist. Now look at the diagram and read what it says.

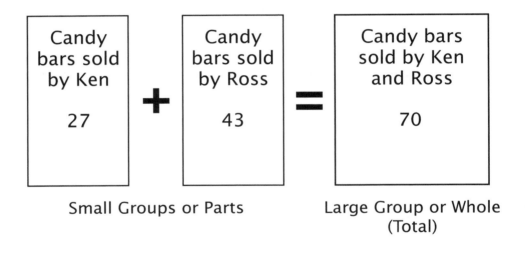

Small Groups or Parts   Large Group or Whole (Total)

Students: The story has two small groups (27 candy bars sold by Ken and 43 candy bars sold by Ross) that combine to make a large group (70 candy bars sold by Ken and Ross).

Teacher: Good. The whole (all candy bars sold) is equal to the sum of the parts (candy bars sold by Ken and candy bars sold by Ross). Does the diagram indicate that the whole is equal to the sum of the parts in this story? How do you know?

Students: Yes, because $70 = 27 + 43$.

Teacher: Great job! What is this problem called? Why?

Students: Group, because two small groups (candy bars sold by Ken and candy bars sold by Ross) combine to make a large group (all candy bars sold). Also, the whole (70) is equal to the sum of the parts (27 + 43 = 70).

Group Story 3

Teacher: (*Use the script as a guideline for mapping information in Group Story 3, and facilitate understanding and reasoning by having frequent student–teacher exchanges. Display Overhead Modeling page for Group Story 3. Have students look at student pages of Group Story 3.*)
  Touch Story 3. What's the first step? What do we need to do? (*Point to first check box under Step 1 on Group Story Checklist.*)

Students: Find the problem type by reading the story and retelling it.

Teacher: Great! (*Read the story aloud, or have a student read it.*)
  "68 students at Hillcrest Elementary took part in the school play. There were 22 third graders, 19 fourth graders, and 27 fifth graders in the school play."
  You read the problem. What must you do next?

Students: Retell the story using own words.

Teacher: Yes. Retell the story in your own words to help you understand it. (*Call on students to retell the story. When they retell the story, remind students to tell what they know in the story.*)

Students: Altogether, 68 students were in the school play; 22 third graders, 19 fourth graders, and 27 fifth graders took part in the school play.

Teacher: Check off the first box under Step 1 on the checklist. (*Point to second check box under Step 1.*) What kind of a problem is this? How do you know?

Students: It is a group problem, because three small groups or parts (third graders, fourth graders, and fifth graders) combine to make a large group or whole (all students).

Teacher: Correct! Check off the second box under Step 1 on the checklist. (*Point to first box under Step 2.*) Now you are ready for Step 2: Organize the information in the problem using the group diagram. To organize the information in a group problem, what must you do first?

Students: Underline the small groups or parts and the large group or whole and write them in the group diagram.

Teacher: How many small groups or parts do we have in this story? What are they?

Students: Three small groups of third graders, fourth graders, and fifth graders.

Teacher: Good, underline them in the story and write <u>third graders in the school play</u> for one of the small groups, <u>fourth graders in the school play</u> for the next small group, and <u>fifth graders in the school play</u> for the other small group in the diagram. (*Direct students to draw* *another box in the diagram for the third small group. Pause for students to write.*) What is the large group or whole?

Students: All students (i.e., third, fourth, and fifth graders) in the school play.

Teacher: Excellent! Underline <u>students</u> in the story and write it for the large group. Check off the first box under Step 2 on the checklist. (*Point to second box under Step 2.*) Now read each sentence, circle the numbers given for the large group and small groups, and write them in for each group. What does the first sentence say?

Students: "68 students at Hillcrest Elementary took part in the school play."

Teacher: Does this sentence tell about the large group or one of the small groups? How do you know?

Students: The large group, because it tells us about *all* the students (third, fourth, and fifth graders) at Hillcrest Elementary who took part in the school play.

Teacher: Great! Circle "68" and write it for the large group amount in the diagram. (*Pause for students to complete.*) What does the next sentence say?

Students: "There were 22 third graders, 19 fourth graders, and 27 fifth graders in the school play."

Teacher: What does this sentence talk about?

Students: Third graders, fourth graders, and fifth graders in the school play.

Teacher: Circle the numbers for each small group and write them in for third graders, fourth graders, and fifth graders in the diagram. (*Pause for students to complete and check their work.*) You circled the numbers for the groups and wrote group names and numbers in the group diagram. Check off the second box under Step 2 on the checklist. Now look at the diagram and read what it says.

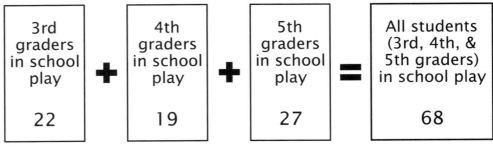

Small Groups or Parts   Large Group or Whole (Total)

Students: There are three small groups of 22 third graders, 19 fourth graders, and 27 fifth graders, and a large group or total of 68 students who took part in the school play.

Teacher: Does this make sense? Why?

Students: This makes sense, because the three small groups together (22 + 19 + 27) equal the large group (68).

Teacher: What is this problem called? Why?

Students: Group, because three small groups combine to make a large group. Also, the whole is equal to the sum of the parts.

Teacher: That's right. Let's review this group problem. The three small groups (22 third graders, 19 fourth graders, and 27 fifth graders in the play) combine to make a large group (all 68 students in the play). Also, the whole (68) is equal to the sum of the parts (22 + 19 + 27 = 68).

(*Pass out Group Schema Worksheet 1.*) Now I want you to do the next four stories on your own. Remember to use the two steps to organize information in stories using the group diagram.

(*Monitor students as they work. Then check the information in the diagrams. Make sure the diagrams are labeled correctly and completely; see below.*)

Group Schema Story 1: "Mrs. Smith's class read 28 books in September, 30 books in October, and 35 books in November. The class read 93 books altogether in the three months."

Group Schema Story 2: "On Tuesday, Michelle rode her bike for 12 miles and walked for 5 miles. Michelle traveled 17 miles on Tuesday."

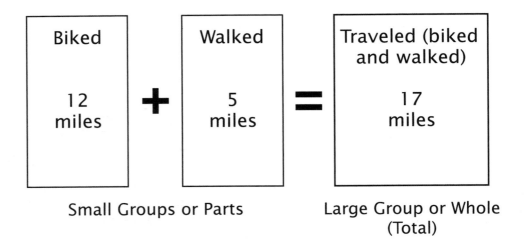

Group Schema Story 3: "Mr. Bradley owns 17 rabbits. 6 of the rabbits are long-eared and 11 are short-eared."

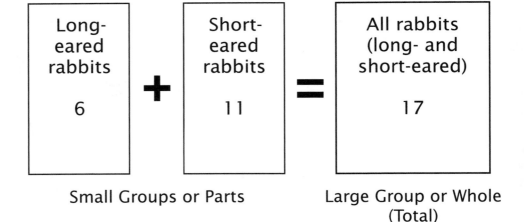

Group Schema Story 4: "Larry and Bart filled 72 buckets of popcorn to sell at a movie. Larry filled 32 buckets of popcorn. Bart filled 40 buckets of popcorn."

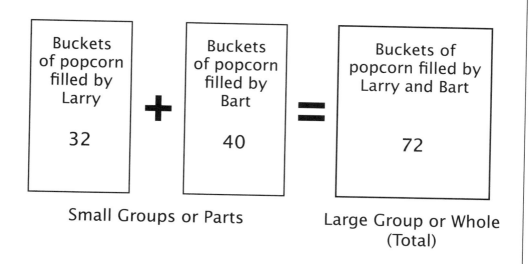

Small Groups or Parts   Large Group or Whole (Total)

Teacher: Today you learned to map information in group story situations onto diagrams. (*Review the main features of a change problem and a group problem.*) Next you will learn to solve group problems. Later you will learn to organize information and solve the compare problem type.

Lesson 7: Problem Solution

Materials Needed

Checklists   Word Problem–Solving Steps (FOPS) poster

                             Group Problem-Solving Checklist (laminated copies for students)

Diagram   Group Problem diagram poster

Overhead Modeling   Lesson 7: Group Problems 1, 2, and 3

Reference Guide   Lesson 7: Group Reference Guide 1

Student Pages   Lesson 7: Group Problems 1, 2, and 3

Teacher: Today we are going to use group diagrams like the ones you learned earlier to solve group word problems. Let's review the group problem. A group problem has two or more small groups or parts that combine to make a large group or whole, and the whole is equal to the sum of the parts. [*Display Word Problem-Solving Steps (FOPS) poster.*] Remember the funny word, FOPS. What are the four steps in FOPS?

Students: F—Find the problem type; O—Organize the information using a diagram; P—Plan to solve the problem; S—Solve the problem.

Group Problem 1

Teacher: (*Display Overhead Modeling page for Group Problem 1, and see Group Reference Guide 1 to set up the problem. Pass out student copies of Group Problems 1, 2, and 3 and Group Problem–Solving Checklist.*)

(*Point to Group Problem–Solving Checklist.*) We will use this checklist that has the same four steps (FOPS) to help us solve group problems.

We are ready for Step 1: Find the problem type. (*Point to first check box on Group Problem–Solving Checklist.*) To find the problem type, I will read the problem and retell it in my own words. Follow along as I read Problem 1. (*Read problem aloud.*)

"There are 75 different flavors of ice cream. Julie's Treats has 35 flavors. How many flavors of ice cream are not at Julie's Treats?"

Now I'll retell the problem in my own words to help me understand it. When I retell, I will ask myself, What do I know in this problem and what am I asked to find out? (*Retell the problem.*)

I know that there are a total of 75 different flavors of ice cream. I also know that Julie's Treats has 35 flavors of ice cream. I don't know the number of flavors of ice cream that are not at Julie's Treats. I need to solve for this amount.

I read the problem and told it in my own words. I will check off the first box under Step 1 on the checklist. (*Point to second check box under*

Step 1.) Now I will ask myself if the problem is a group problem. Why do you think this is a group problem? What are the small groups or parts that combine to make the large group or whole?

Students: The two small groups or parts are "flavors of ice cream at Julie's Treats" and "flavors not at Julie's Treats." The large group or whole is "all the different flavors of ice cream."

Teacher: Let's check off the second box under Step 1. Now I am ready for Step 2: Organize the information in the problem using the group diagram. (*Display Group Problem diagram poster.*)

Group Problem

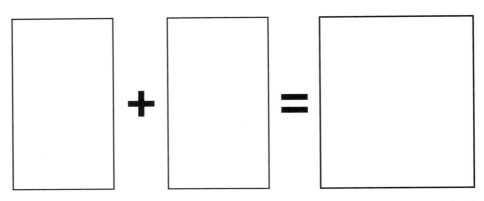

Small Groups or Parts   Large Group or Whole (Total)

(*Point to first box under Step 2.*) To organize the information in a group diagram, we first underline the group names and write them in the diagram. In this problem, we have two small groups or parts. What are the two small groups in this problem?

Students: Flavors of ice cream at Julie's Treats and flavors of ice cream not at Julie's Treats.

Teacher: Good, underline <u>Julie's Treats flavors</u> in the problem and write "flavors of ice cream at Julie's Treats" for one of the small groups in the diagram. Underline <u>flavors of ice cream not at Julie's Treats</u> and write it for the other small group in the diagram. What is the large group or whole?

Students: All flavors of ice cream.

Teacher: Excellent! Underline <u>different flavors of ice cream</u> in the problem and write "flavors of ice cream at Julie's Treats and not at Julie's Treats" for the large group. (*Pause for students to write.*) Let's check off the first box under Step 2 on the checklist. (*Point to second box under Step 2.*) Now let's read the problem and circle the numbers given for the small groups and the large group. The first sentence says, "There are 75 different flavors of ice cream." Does this sentence tell about the large group or one of the small groups? How do you know?

Students: The large group, because it tells us about all the different flavors of ice cream at and not at Julie's Treats.

Teacher: Do we know this amount? What is it?

Students: Yes. It is 75.

Teacher: Circle "75" and write it for the large group. The next sentence says, "Julie's Treats has 35 flavors." Which one of the two small groups does this sentence describe?

Students: Flavors of ice cream at Julie's Treats.

Teacher: Good. Do we know how many flavors of ice cream there are at Julie's Treats from this sentence?

Students: Yes. There are 35 flavors.

Teacher: Circle "35" and write it for the small group (i.e., flavors of ice cream at Julie's Treats). The last sentence is a question, because it asks, "How many flavors of ice cream are not at Julie's Treats?" (*Point to third box under Step 2 in checklist.*) This sentence refers to the other small group amount (i.e., flavors of ice cream not at Julie's Treats) that we need to solve for in this problem. I don't know this amount, so I will write a "?" We circled the numbers for the groups and wrote them in the diagram, and we wrote a "?" for the small group amount we need to solve for. Let's check off the second and third boxes under Step 2 in the checklist.

Now let's look at the diagram and read what it says. (There are 35 flavors of ice cream at Julie's Treats and a total of 75 different flavors of ice cream at and not at Julie's Treats. I need to find out the number of flavors of ice cream not at Julie's Treats). What must you solve for in this problem? (Is it a small group or the large group amount?)

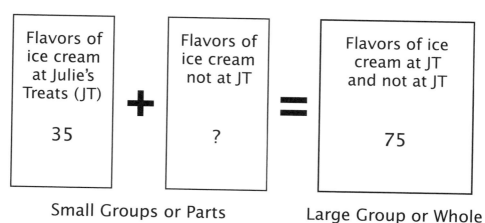

Students: The small group amount.

Teacher: Now for Step 3: Plan to solve the problem. (*Point to first box under Step 3 on checklist.*) To plan how to solve the problem, we first decide whether to add or subtract. To do that, I must ask if the total or whole is given. In a group problem, the total or whole is the large group quantity. Remember, if the total or whole is given, subtract; if the total or whole is not given, add. From this diagram, I know the large group quantity or the whole. Do we add or subtract to find the part (small group amount)?

Students: Subtract.

Teacher: Good. Let's check off the first box under Step 3 on the checklist. (*Point to second box under Step 3 on checklist.*) Next we write the math sentence. Remember, when we subtract, we start with the total or whole and subtract the other number. Also, remember to line up the numbers correctly. So the math sentence is 75 − 35 = ?. Let's write the math sentence and check off the second box under Step 3 of the checklist. (*Pause for students to complete.*)

Now we are ready for Step 4: Solve the problem. (*Point to first box under Step 4 on checklist.*) What does 75 − 35 =? (*Pause for students to solve the problem.*)

Students: 40.

Teacher: Check off the first box under Step 4 on the checklist. (*Point to second box under Step 4.*) Now write "40" for the "?" in the diagram and write the complete answer on the answer line. The complete answer is the number and the label. What is the complete answer to this group problem?

Students: 40 flavors of ice cream not at Julie's Treats.

Teacher: Good. Write "40 flavors of ice cream not at Julie's Treats" on the answer line. (*Pause for students to write answer.*) Let's check off the second box under Step 4 on the checklist. (*Point to third box under Step 4 in checklist.*) We are now ready to check the answer. Does "40" seem right? Yes, because 40 flavors of ice cream not at Julie's Treats is the small group and is less than the large group (75 flavors of ice cream). We can also check by adding to see that the whole is equal to the sum of the parts: 75 = 35 + 40. (*Check off third box under Step 4.*)

Let's review this problem. (*Model writing the explanation for how the problem was solved here; see Group Reference Guide 1.*)

Let's review. What's this problem called? Why?

Students: Group, because two small groups or parts combine to make a large group or whole.

Teacher: Yes. Also, the whole (75) is equal to the sum of the parts (35 + 40 = 75).

Group Problem 2

Teacher: (*Display Overhead Modeling page for Group Problem 2. Have students look at student pages of Group Problem 2.*)
Touch Group Problem 2. What's the first step? (*Point to checklist.*)

Students: Find the problem type.

Teacher: Right. (*Point to first check box on Group Problem–Solving Checklist.*) To find the problem type, I will read the problem and retell it in my own words. (*Read the problem aloud.*)

"Use the information in the graph below to solve Group Problem 2."

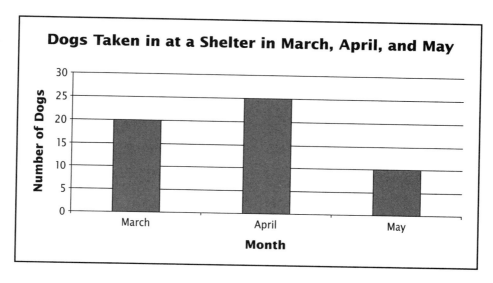

"How many dogs did the shelter take in during March, April, and May?"
I read the problem. What must I do next?

Students: Retell the problem using own words.

Teacher: Yes, I will retell the problem in my own words to help me understand it. When I retell, I will ask myself, What do I know in this problem and what am I asked to find out? (*Retell the problem.*)
I know that the shelter took in 20 dogs in March, 25 dogs in April, and 10 dogs in May. I don't know how many dogs were taken in at the shelter across the three months.
I read the problem and told it in my own words. I will check off the first box under Step 1 on the checklist. (*Point to second check box under Step 1.*) Now I will ask myself if the problem is a group problem. Why do you think this is a group problem? What are the small groups or parts that combine to make the large group or whole?

Students: The three small groups are the number of dogs taken in at the shelter in each of the three months, March, April, and May. The large group or whole is all the dogs taken in the shelter during all three months combined.

Teacher: Let's check off the second box under Step 1. Now I am ready for Step 2: Organize the information in the problem using the group diagram. (*Display Group Problem diagram poster.*)

Group Problem

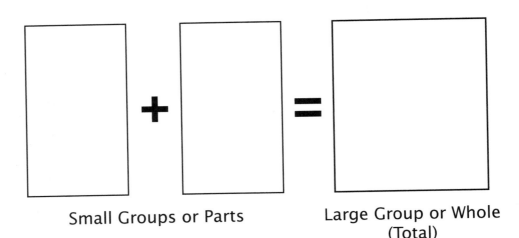

Small Groups or Parts   Large Group or Whole (Total)

(*Point to first box under Step 2.*) To organize the information in a group diagram, we first underline the group names and write them in the diagram. In this problem, the data for the groups are given in the graph. You can see from the graph that we have three small groups or parts. What are the three small groups in this problem?

Students: Dogs taken in the shelter in March, dogs taken in the shelter in April, and dogs taken in the shelter in May.

Teacher: Good. Let's draw one more box to the left for another small group in the diagram. First underline <u>March</u> in the graph and write "dogs taken in shelter in March" for one of the small groups. Next underline <u>April</u> in the graph and write "dogs taken in shelter in April" in the diagram for the second small group. Then underline <u>May</u> in the graph and write "dogs taken in shelter in May" in the diagram for the third small group. (*Pause for students to write.*) What is the large group or whole?

Students: All dogs taken in the shelter in March, April, and May.

Teacher: Excellent! Underline <u>dogs taken in shelter in March, April, and May</u> in the question and write it for the large group. (*Pause for students to write.*) Let's check off the first box under Step 2 on the checklist. (*Point to second box under Step 2.*)

Now let's read the problem and circle the numbers given for the small groups and the large group. The sentence asks, "How many dogs did the shelter take in during March, April, and May?" (*Point to third box under Step 2 in checklist.*) Does this sentence ask about the large group or one of the small groups? How do you know?

Students: Large group, because it talks about all dogs taken in the shelter during the months of March, April, and May combined.

Teacher: Do we know this amount?

Students: No.

Teacher: Let's write a "?" in our diagram for the large group because this is what we need to solve for in this problem. What must I solve for in this problem (the small group or large group amount)?

Students: The large group.

Teacher: Good. Next we need to read the graph to find the numbers for the small groups in this problem. From this graph, we know that 20 dogs were taken in the shelter in March. (*Model how to read information from a graph. Have students circle the number corresponding to each small group in the graph.*) Write 20 for "March" (small group) in the diagram. The next part of the graph shows the number of dogs taken in the shelter in April, which is another small group. How many dogs were taken in the shelter in April?

Students: 25.

Teacher: Circle "25" and write in "25" for "April" (small group) in the diagram. The next part of the graph shows the number of dogs taken in the shelter in May, which is another small group. How many dogs were taken in the shelter in May?

Students: 10.

Teacher: Circle "10" and write "10" for "May" (small group) in the diagram. We circled the numbers for the groups and wrote group names and numbers in the diagram. We also wrote a "?" for the large group amount we need to solve for. Let's check off the second and third boxes under Step 2 in the checklist. (*Pause for students to complete.*) Now let's look at the diagram and read what it says. (There were 20 dogs taken in the shelter in March, 25 in April, and 10 in May. We need to find out how many dogs in all were taken in the shelter in the three months?) What must you solve for in this problem? (The small group or large group amount?)

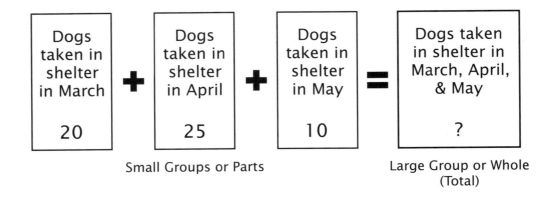

Small Groups or Parts   Large Group or Whole (Total)

Students: Large group amount.

Teacher: Now for Step 3: Plan to solve the problem. (*Point to first box under Step 3 on checklist.*) To plan how to solve the problem, we first decide whether to add or subtract. To do that, I must ask if the total or whole is given. In a group problem, the total or whole is the large group quantity. From this diagram, I don't know the large group quantity or the whole. If the whole (total) is not given, do we add or subtract?

Students: Add.

Teacher: Good. Let's check off the first box under Step 3 on the checklist. (*Point to second box under Step 3.*) Next we write the math sentence. Remember, we have to add and we have to find the whole or large group amount. Also, remember to line up the numbers correctly. So the math sentence is 20 + 25 + 10 = ?. Let's write the math sentence and check off the second box under Step 3 of the checklist. (*Pause for students to complete.*) Now we are ready for Step 4: Solve the problem. What does 20 + 25 + 10 =? (*Pause for students to solve the problem.*)

Students: 55.

Teacher: Check off the first box under Step 4 on the checklist. (*Point to second box under Step 4.*) Now write 55 for the "?" in the diagram and write the complete answer on the answer line. The complete answer is the number and the label. What is the complete answer to this group problem?

Students: 55 dogs taken in the shelter in March, April, and May.

Teacher: Good. I'll write "55 dogs taken in the shelter in March, April, and May" on the answer line. (*Pause for students to write the answer.*) Let's check off the second box under Step 4 on the checklist. We are now ready to check the answer. (*Point to third box under Step 4 in checklist.*) Does 55 seem right?

Students: Yes, because the large group number (all dogs taken in the shelter in March, April, and May) is more than each of the small group numbers.

Teacher: (*Check off third box under Step 4. Guide students to write the explanation for solving the problem on their worksheet; see Group Reference Guide 1.*) Excellent. Let's review. What's this problem called? Why?

Students: Group, because three small groups or parts combine to make a large group or whole.

Teacher: Good. Also, the whole (55) is equal to the sum of the parts (20 + 25 + 10 = 55).

Group Problem 3

Teacher: (*Use the script as a guideline for solving Group Problem 3, and facilitate the problem-solving process by having frequent student–teacher exchanges. Display Overhead Modeling page for Group Problem 3. Have students look at student pages of Group Problem 3.*) Touch Problem 3. What's the first step?

Students: Find the problem type.

Teacher: Right. (*Point to first check box on Group Problem–Solving Checklist.*) To find the problem type, what must you do?

Students: Read the problem and retell it in own words.

Teacher: Good. Read the problem aloud.

Students: "In a poll, 98 people were asked what their favorite flower is, and 45 named roses. How many named another kind of flower?"

Teacher: You read the problem. What must you do next?

Students: Retell the problem using my own words.

Teacher: Yes. Now retell the problem in your own words to help you understand it. (*Call on students to read and retell the problem. When they retell the problem, remind students to tell what they know in the problem and what they are asked to find out.*)

Students: I know that a total of 98 people answered the poll about their favorite flower. I know that 45 of them named roses as their favorite flower. I don't know how many named another kind of flower as their favorite flower.

Teacher: Check off the first box under Step 1 on the checklist. (*Point to second check box under Step 1.*) What kind of a problem is this? How do you know?

Students: It is a group problem, because there is a large group (i.e., all the people polled about their favorite flower), a small group (i.e., people who named roses as their favorite flower), and another small group (i.e., people who named another kind of flower as their favorite flower).

Teacher: Check off the second box under Step 1 on the checklist. Now you are ready for Step 2: Organize the information in the problem using the group diagram. (*Point to first box under Step 2 of checklist.*) To organize the information in a group problem, what must you do first?

Students: Underline the names of the small groups and large group.

Teacher: Good. In this problem, what are the two small groups?

Students: People who named roses and people who named another kind of flower.

Teacher: Good, underline <u>named roses</u> and write "people who named roses" for one of the small groups in the diagram. Underline <u>named another kind of flower</u> and write "people who named another kind of flower" for the other small group in the diagram. What is the large group or whole?

Students: All people who named roses and another kind of flower.

Teacher: Excellent! Underline <u>people asked what their favorite flower is</u> in the problem and write in "people who named roses and another kind of flower" for the large group. (*Pause for students to write.*) Let's check off the first box under Step 2 on the checklist. (*Point to second box under Step 2.*) What do you do next?

Students: Read the problem and circle the numbers given for the small groups and the large group.

Teacher: The first sentence says, "In a poll, 98 people were asked what their favorite flower is, and 45 named roses." What does this sentence tell about? (The large group or the small groups?)

Students: The large group of 98 people who named roses and another kind of flower and one of the small groups (45 people who named roses as their favorite flower).

Teacher: Good, circle "98" and write it for the large group. Also circle "45" and write it for one of the small groups (i.e., people who named roses). The last sentence is the question. (*Point to third box under Step 2 in checklist.*) It asks, "How many named another kind of flower?" What does this question ask us to solve for? How do you know?

Students: It asks us to solve for the other small group amount (i.e., people who named another kind of flower).

Teacher: We don't know this amount, so let's write a "?" in the box. We circled numbers for the groups and wrote them in the diagram. We also wrote a "?" for the small group amount that we need to solve for. Let's check off the second and third boxes under Step 2 in the checklist. Now look at the diagram and read what it says.

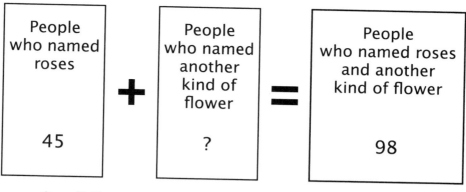

Small Groups or Parts   Large Group or Whole
                                    (Total)

Students: 45 people named roses, and a total of 98 people named roses and another kind of flower. I need to find out the number of people who named another kind of flower.

Teacher: What must you solve for in this problem? (*Is it a small group or the large group amount?*)

Students: A small group amount.

Teacher: Good. Now for Step 3: Plan to solve the problem. To plan how to solve the problem, what must you do? (*Point to first box under Step 3 on checklist.*)

Students: Decide whether to add or subtract.

Teacher: To figure out whether to add or subtract to solve this problem, what should you ask yourself?

Students: If the total or whole is given or not given.

Teacher: Right! In a group problem, the whole or total is the large group quantity. From this diagram, do you know the whole or total?

Students: Yes.

Teacher: If the total is given, do you add or subtract to find the part (small group amount)?

Students: Subtract.

Teacher: Good. Check off the first box under Step 3 on the checklist. What do you do next? (*Point to second box under Step 3 on checklist.*)

Students: Write the math sentence.

Teacher: Remember, you have to subtract. So start with the whole or total and subtract the other number. Also, remember to line up the numbers correctly. What is the math sentence?

Students: 98 − 45 = ?.

Teacher: Write it and check off the second box under Step 3 on the checklist. (*Pause for students to complete.*) Now you are ready for Step 4: Solve the problem. What do you do first?

Students: Solve for 98 − 45.

Teacher: What does 98 − 45 =? (*Pause for students to solve the problem.*)

Students: 53.

Teacher: Check off the first box under Step 4 on the checklist. What do you do next? (*Point to second box under Step 4.*)

Students: Write the complete answer.

Teacher: The complete answer is the number and the label. What is the complete answer to this group problem?

Students: 53 people named another kind of flower.

Teacher: Good. Write "53" for the "?" in the diagram and write "53 people named another kind of flower" on the answer line. (*Pause for students to write the answer.*) Let's check off the second box under Step 4 on the checklist. (*Point to third box under Step 4.*) What do you do next?

Students: Check the answer.

Teacher: Does 53 seem right?

Students: Yes, because 53 is the number of people who named another kind of flower. This is the small group and is less than the large group amount of 98 people who named both roses and another kind of flower.

Teacher: We can also check by adding to see that the whole (98) is equal to the sum of the parts (i.e., 45 + 53 = 98). (*Check off third box under Step 4.*) Let's review this problem. What's this problem called? Why?

Students: Group, because two small groups combine to make a large group.

Teacher: Yes. Also, the whole (98) is equal to the sum of the parts (45 + 53 = 98). You did an excellent job working hard. Tomorrow, we will practice more group problems.

Lesson 8: Problem Solution

Materials Needed

Answer Sheet for Paired-Learning Tasks	Lesson 8: Group Answer Sheet 1
Checklists	Word Problem–Solving Steps (FOPS) poster
	Group Problem–Solving Checklist (laminated copies for students)
Diagram	Group Problem diagram
Overhead Modeling	Lesson 8: Group Problem 4
Student Pages	Lesson 8: Group Problem 4
	Lesson 8: Group Worksheet 1

Group Problem 4

Teacher: [*Display Word Problem–Solving Steps (FOPS) poster. Ask students to read each step on the poster, which they will use to solve addition and subtraction word problems. Display Overhead Modeling page for Group Problem 4. Have students look at student pages of Group Problem 4.*]
   Touch Problem 4. (*Point to Group Problem–Solving Checklist.*) Remember, we will use this checklist that has the same four steps (FOPS) to help us solve group problems. What's the first step?

Students: Find the problem type.

Teacher: Right. (*Point to first check box on checklist.*) To find the problem type, what must you do?

Students: Read the problem and retell it in your own words.

Teacher: Good. Read the problem aloud.

Students: "A photographer spent a total of $95 to photograph the Grand Canyon with his dad's camera. He spent $50 on a tripod, $30 on batteries, and some on film. How much did he spend on film?"

Teacher: You read the problem. What must you do next?

Students: Retell the problem using my own words.

Teacher: Yes. Now retell the problem in your own words to help you understand it. (*Call on students to read and retell the problem. When they retell the problem, remind students to tell what they know in the problem and what they are asked to find out.*)

Students: I know that the photographer spent a total of $95 to photograph the Grand Canyon. I also know that he spent $50 on a tripod and $30 on batteries. I don't know how much he spent on film, which I must solve for in this problem.

Teacher: Good. Check off the first box under Step 1 on the checklist. (*Point to second check box under Step 1.*) What kind of problem is this? How do you know?

Students: Group, because there is a large group (i.e., the tripod, batteries, and film together, or camera equipment). There are three small groups: a tripod, batteries, and film.

Teacher: Check off the second box under Step 1 on the checklist. Now you are ready for Step 2: Organize the information in the problem using the group diagram. To organize the information, what must you do first?

Students: Underline the information about the small groups and large group and write the group names in the diagram.

Teacher: Great! In this problem, what are the three small groups?

Students: The tripod, batteries, and film.

Teacher: Good, draw one more box for the third group or part in the diagram. Underline <u>tripod</u> and write it for one of the small groups in the diagram. Underline <u>batteries</u> and write it for the second small group in the diagram. Underline <u>film</u> and write it for the third small group in the diagram. What is the large group or whole?

Students: The tripod, batteries, and film, or all camera equipment.

Teacher: Excellent! Write "tripod, batteries, and film" for the large group. Because the small groups and the large group all talk about dollars, let's write a dollar sign in the diagram for all groups. (*Pause for students to write.*) Let's check off the first box under Step 2 on the checklist. (*Point to second box under Step 2.*) What do you do next?

Students: Read the problem and circle the numbers given for the small groups and the large group.

Teacher: The first sentence says, "A photographer spent a total of $95 to photograph the Grand Canyon with his dad's camera." Does this sentence tell about the large group or one of the small groups? How do you know?

Students: The large group, because it tells about the total money spent to photograph the Grand Canyon.

Teacher: Do we know this amount? What is it?

Students: $95.

Teacher: Good, circle "$95" and write it for the large group (i.e., tripod, batteries, and film). The next sentence says, "He spent $50 on a tripod,

$30 on batteries, and some on film." This sentence tells about the small groups. Do we know the amount for the tripod? What is it?

Students: $50.

Teacher: Good, circle "$50" and write it for the tripod (i.e., that small group). Do we know the amount for the batteries? What is it?

Students: $30.

Teacher: Good, circle "$30" and write it for the batteries. Do we know the amount for the film?

Students: No.

Teacher: That's right! We don't know the amount spent on film. The last sentence is the question. (*Point to third box under Step 2 in checklist.*) It asks, "How much did he spend on film?" What does this question ask us to solve for? How do you know?

Students: It asks us to solve for the small group amount, which is the money spent on film.

Teacher: We don't know this amount, so let's write a "?" in that box. We circled numbers for the groups and wrote them in the diagram. We also wrote a "?" for the small group amount that we need to solve for. Let's check off the second and third boxes under Step 2 in the checklist. Now look at the diagram and read what it says.

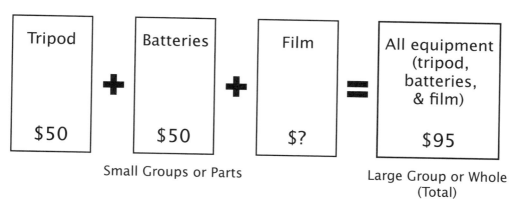

Small Groups or Parts   Large Group or Whole (Total)

Students: A total of $95 was spent on camera equipment (i.e., the tripod, batteries, and film). The tripod cost $50 and the batteries cost $30. I need to find out how much was spent on film.

Teacher: What must you solve for in this problem? (Is it a small group or the large group amount?)

Students: A small group amount.

Teacher: Now for Step 3: Plan to solve the problem. To plan how to solve the problem, what must you do? (*Point to first box under Step 3 on checklist.*)

Students: Decide whether to add or subtract.

Teacher: To figure out whether to add or subtract to solve this problem, what should you ask yourself?

Students: If the total or whole is given or not given.

Teacher: Right! In a group problem, the whole or total is the large group quantity. From this diagram, do you know the whole or total?

Students: Yes.

Teacher: If the whole or total is given, do you add or subtract to find the small group amount (part)?

Students: Subtract.

Teacher: Good. Check off the first box under Step 3 on the checklist. What do you do next? (*Point to second box under Step 3.*)

Students: Write the math sentence.

Teacher: Remember, you have to subtract. Because we have three small groups or parts, we need to first add the two given small parts (i.e., $50 and $30) and then subtract this amount from the total or large group amount ($95). What is the first math sentence?

Students: $50 + $30 = $?.

Teacher: What is $50 + $30?

Students: $80.

Teacher: Great! Now subtract this amount ($80) from the total or $95. Remember to line up your numbers correctly. What is the second math sentence?

Students: $95 − $80 = ?.

Teacher: Write it and check off the second box under Step 3 on the checklist. (*Pause for students to complete.*) Now you are ready for Step 4: Solve the problem. What do you do first?

Students: Solve for 95 − 80.

Teacher: What does 95 − 80 =? (*Pause for students to solve the problem.*)

Students: $15.

Teacher: Check off the first box under Step 4 on the checklist. What do you do next? (*Point to second box under Step 4.*)

Students: Write the complete answer.

Teacher: The complete answer is the number and the label. What is the complete answer to this group problem?

Students: $15 was spent on film.

Teacher: Good. Write "15" for the "?" in the diagram and write "$15 spent on film" on the answer line. (*Pause for students to write the answer.*) Let's check off the second box under Step 4 on the checklist. We are now ready to check the answer. (*Point to third box under Step 4 on checklist.*) What do you do next?

Students: Check the answer.

Teacher: Does $15 seem right?

Students: Yes, because $15 is the amount spent on film. This is the small group and is less than the large group amount of $95.

Teacher: We can also check by adding the small group amounts: $50 + $30 + $15 = $95. (*Check off third box under Step 4.*) Is the whole equal to the sum in this problem? How do you know?

Students: Yes, $95 = $50 + $30 + $15.

Teacher: Good. Let's review this problem. What's this problem called? Why?

Students: Group, because three small groups combine to make a large group. Also, the whole ($95) is equal to the sum of the parts ($50 + $30 + $15 = $95).

Teacher: (*Pass out Group Worksheet 1.*) Now I want you to do Problem 1 on this worksheet with your partner.

(*Ask students to* think, plan, *and* share *with partners to solve Group Worksheet 1, Problem 1; see Guide to Paired Learning in the Introduction.*)

Group Worksheet 1, Problem 1: "Three buses took 90 students on a field trip. One bus carried 35 students, another bus carried 28 students, and the third bus carried the remaining students. How many students were in the third bus?"

Use the four steps to solve the group problem.
(*Monitor students as they work. Have students check their answers using Group Answer Sheet 1. Make sure the diagram is labeled correctly, the math sentence is written and worked out correctly, the written explanation is complete, and the complete answer is written on the answer line; see below.*)

Addition and Subtraction

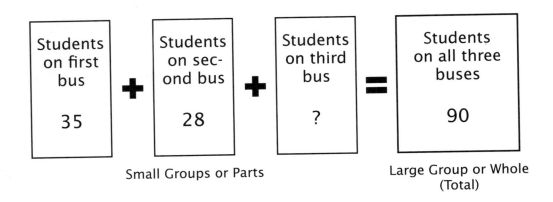

Small Groups or Parts   Large Group or Whole (Total)

Answer: 27 students were in the third bus

Teacher: Now I want you to do the next two problems on your own. Remember to use the four steps to solve these problems.

Group Worksheet 1, Problem 2: "A rose plant has 12 flowers and 10 buds. If 5 buds open, how many buds are left unopened?"

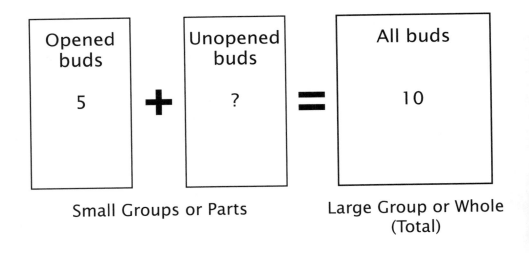

Small Groups or Parts   Large Group or Whole (Total)

Answer: 5 unopened buds

Group Worksheet 1, Problem 3: "In a park, there were 8 ducks in the pond and 5 ducks outside the pond. How many ducks were there in all?"

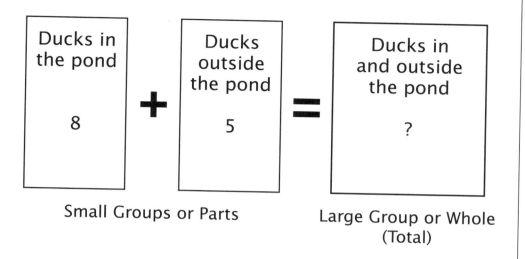

Small Groups or Parts   Large Group or Whole (Total)

Answer: 13 ducks in all

Teacher: Great job working hard. Tomorrow we will practice more group problems.

Lesson 9: Problem Solution

Materials Needed

Answer Sheet for Paired-Learning Tasks	Lesson 9: Group Answer Sheet 2
Checklists	Word Problem–Solving Steps (FOPS) poster
	Group Problem–Solving Checklist
Diagram	Group Problem diagram poster
Overhead Modeling	Lesson 9: Group Worksheet 2, Problem 1
Student Pages	Lesson 9: Group Worksheet 2

Teacher: (*Pass out Group Worksheet 2. Display Overhead Modeling page for Group Worksheet 2, Problem 1.*)

Follow along as I read this problem. (*Use guided practice to have students complete Group Worksheet 2, Problem 1. Read Problem 1 aloud.*)

"Use data from the table below to solve Group Worksheet 2, Problem 1."

Animals at Green Acres Farm

Animals	Number of Animals
Chickens	20
Cows	32
Ducks	23
Horses	39
Pigs	21
Sheep	19

Group Worksheet 2, Problem 1: "Green Acres Farm has many animals. How many chickens, ducks, and pigs live at Green Acres Farm?"

Answer: 64 chickens, ducks, and pigs

Teacher: Now I want you to do the next problem with your partner. (*Ask students to think, plan, and share with partners to solve Group Worksheet 2, Problem 2; see Guide to Paired Learning section in the Introduction.*)

Group Worksheet 2, Problem 2: "Farmer Jake has 88 animals on his farm. He only has horses and goats. There are 49 horses on the farm. How many goats are on the farm?"

Use the four steps to solve this group problem.

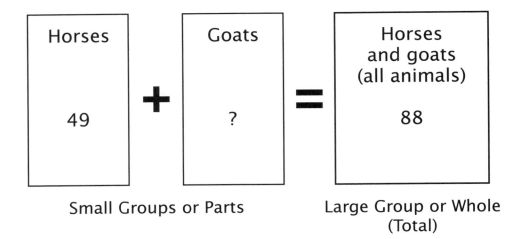

Small Groups or Parts   Large Group or Whole (Total)

Answer: 39 goats

(*Monitor students as they work. Have students check their answers using Group Answer Sheet 2. Make sure the diagram is labeled correctly, the math sentence is written and worked out correctly, the written explanation is complete, and the complete answer is written on the answer line; see below.*)

Teacher: Now I want you to do the next four problems on your own. Remember to use the four steps to solve the problems.
"Use data from the table below to solve Group Worksheet 2, Problem 3."

Items Bought at the Market	Prices of the Items
Milk	$2.09
Butter	$1.62
Cheese	$1.89
Bread	$1.78

Group Worksheet 2, Problem 3: "You go to the market. You decide to buy milk and bread. How much will both items cost?"

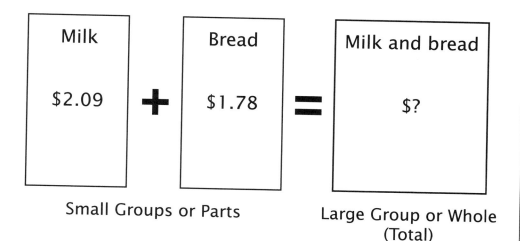

Answer: Milk and bread cost $3.87

Group Worksheet 2, Problem 4: "Last week, 79 children signed up for softball. 52 of the children have received their team shirts. How many children still need to receive their shirts?"

Answer: 27 children still need to receive team shirts

Group Worksheet 2, Problem 5: "At Joe's birthday party, you got 15 party favors. You have 8 games, 4 balloons, and some whistles. How many whistles did you get?"

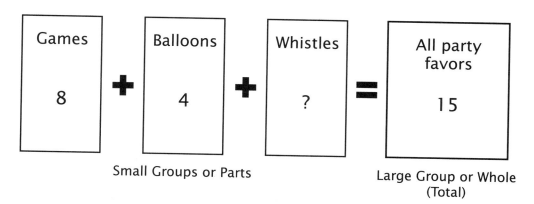

Answer: 3 whistles

Group Worksheet 2, Problem 6: "Phyllis took 18 pictures. 9 are of Niagara Falls. How many pictures are not of Niagara Falls?"

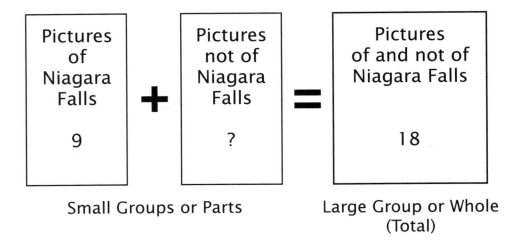

Small Groups or Parts   Large Group or Whole (Total)

Answer: 9 pictures are not of Niagara Falls

Teacher: (*Monitor students as they work. After about 15 to 20 minutes, go over the answers. Make sure the diagrams are labeled correctly, the math sentences are written and worked out correctly, and the complete answers are written on the answer line.*)

Great job working hard. Tomorrow we will solve some more group problems and then get ready to learn the next problem type, a compare problem.

Lesson 10: Problem Solution

Materials Needed

Checklists	Word Problem–Solving Steps (FOPS) poster
	Group Problem–Solving Checklist (laminated copies for students)
Overhead Modeling	Lesson 10: Group Worksheet 3, Problems 1 and 2
Reference Guide	Lesson 10: Group Reference Guide 2
Student Pages	Lesson 10: Group Worksheet 3

Teacher: (*Pass out Group Worksheet 3. Display Overhead Modeling page for Group Worksheet 3, Problem 1.*)

You learned to solve group problems using diagrams. Now you will solve the problems on this worksheet using your own diagrams. This worksheet does not have diagrams. Remember to use the four steps to solve the problems on the worksheet. (*Note: Discuss how students can generate a diagram that is more efficient than the Group Problem diagram used in the previous lessons, and have them practice solving the problems using the diagram they generate. Also, encourage them to use the Group Problem–Solving Checklist only as needed.*)

(*Use guided practice to complete Group Worksheet 3, Problems 1 and 2, using own diagrams.*)

(*Monitor students as they work. After about 15 to 20 minutes, go over the answers. Make sure the diagrams are labeled correctly, the math sentences are worked out correctly, and the complete answers are written on the answer line; see below.*)

Group Worksheet 3, Problem 1: "During a tour of the Botanical Gardens, Tom saw a patch of 35 plants. Insects were resting on 8 of the plants. How many plants did not have insects on them?"

Plants with insects resting		Plants with no insects resting		All plants
8	+	?	=	35
SG		SG		LG

Answer: 27 plants did not have insects on them

"Use the graph below to answer Group Worksheet 3, Problem 2."

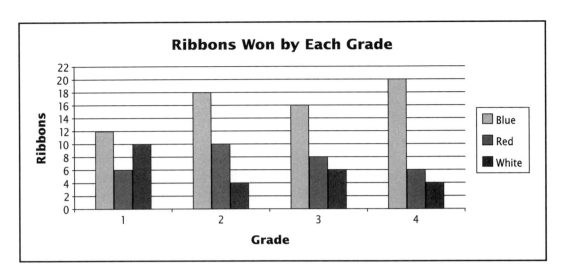

Group Worksheet 3, Problem 2: "Moser Elementary School held a science fair. How many red ribbons were awarded at the science fair?" (*See Group Reference Guide 2.*)

Red ribbons won by Grade 1		Red ribbons won by Grade 2		Red ribbons won by Grade 3		Red ribbons won by Grade 4		All red ribbons won
6	+	10	+	8	+	6	=	?
SG		SG		SG		SG		LG

Answer: 30 red ribbons were awarded at the science fair

Teacher: Now I want you to do the next three problems on your own. (*Have students write the explanation for at least one of the three problems.*) Remember to use the four steps to solve Problems 3 through 5 on the worksheet.

Group Worksheet 3, Problem 3: "A new baseball bat costs $50. A new baseball cap costs $10. How much would it cost to buy the baseball bat and cap?"

New baseball bat		New baseball cap		New baseball bat and cap
$50	+	$10	=	$?
SG		SG		LG

Answer: $60 for the new baseball bat and cap

Group Worksheet 3, Problem 4: "At the flower show, one display won a blue ribbon. It had 28 flowers. 7 of the flowers were white roses. How many of the flowers were not white roses?"

White roses		Not white roses		All flowers
$\underline{7}$	+	$\underline{?}$	=	$\underline{28}$
SG		SG		LG

Answer: 21 not white roses

"Use the information from the bar graph below to solve Group Worksheet 3, Problem 5."

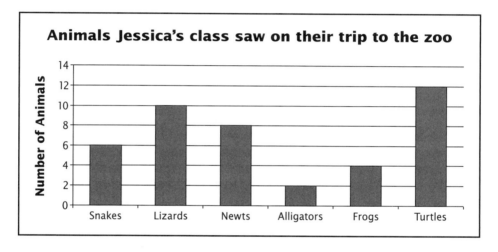

Group Worksheet 3, Problem 5: If the total number of frogs at the zoo was 14, how many frogs did Jessica's class not see on their trip to the zoo?

Frogs seen		Frogs not seen		All frogs at the zoo
$\underline{4}$	+	$\underline{?}$	=	$\underline{14}$
SG		SG		LG

Answer: 10 frogs not seen at the zoo

Teacher: Great job working hard. You learned to solve group word problems using your own diagrams. Next you will learn to map information in compare word problems onto diagrams and solve them.

Unit 3

Compare Problems

Lesson 11: Problem Schema

Materials Needed

Checklist	Compare Story Checklist (laminated copies for students)
Diagram	Compare Problem diagram poster
Overhead Modeling	Lesson 11: Compare Stories 1, 2, and 3
Student Pages	Lesson 11: Change and Group Review Worksheet 1
	Lesson 11: Compare Stories 1, 2, and 3
	Lesson 11: Compare Schema Worksheet 1

Teacher: (*Pass out Change and Group Review Worksheet 1. Review solving change and group problems using Problems 1 through 4 on the worksheet. After the review, proceed with the compare lesson.*)
   Now we will learn to identify and organize another type of addition and subtraction problem called "compare" so that we can later solve it. A compare problem tells about the sameness or difference between two things. More often, the comparison tells about a difference. In a compare problem, two objects, persons, or things are compared using a common unit (e.g., age, height). The comparison sentence in the problem helps us to find the problem type. How do I know if it is a comparison sentence? Compare words, such as *more than, less than, fewer than, younger than, older than, taller than,* and *shorter than,* can help us find the comparison sentence. For example, "Danny has 12 toys. John has 24 *more toys than* Danny." Which sentence is the comparison sentence and why?

Students: "John has 24 more toys than Danny" is the comparison sentence, because the compare words *more than* tell about a comparison.

Teacher: The comparison sentence also helps us figure out the two things that are compared in the problem. For example, "John has 24 more toys than Danny." Who are the two people compared in this problem? What are they compared on?

Students: John and Danny are compared on the number of toys they have.

Teacher: The comparison sentence also tells us which amount is the bigger number. Does John or Danny have more toys?

Students: John.

Teacher: Right, the number of toys John has is the bigger number. The comparison sentence also tells us the difference between the two things being compared. What is the difference amount between the number of toys that John has and the number that Danny has?

Students: 24 toys.

Teacher: Excellent! (*Present a couple more examples orally to help students understand a compare problem type. You may ask students to give the ages of family members to make a chart and help them compare their ages.*)

(*Display Compare Story Checklist.*) Here are two steps we will use to organize information in a compare story. (*Point to each step on Compare Story Checklist and read each one.*)

Compare Story 1

Teacher: Let's use these two steps to do an example. Look at this story. (*Display Overhead Modeling page of Compare Story 1. Pass out student copies of Compare Stories 1, 2, and 3 and Compare Story Checklist. Point to first check box on Compare Story Checklist.*) Now we are ready for Step 1: Find the problem type. To find the problem type, I will read the story and retell it in my own words. (*Read problem aloud.*)

"Joe is 15 years old. He is 8 years older than Jill. Jill is 7 years old."

Now I'll retell the problem in my own words to help me understand it. When I retell, I will ask myself, What do I know in this story? (*Retell the problem.*)

Joe is 8 years older than Jill. Jill is 7 years old and Joe is 15 years old. I read the story and told it in my own words. Let's check off the first box under Step 1 on the checklist. (*Point to second check box under Step 1.*)

Now I will ask myself if the story is a compare problem type. How do I know it is a compare problem? What is this story comparing? The compare words *older than* in the comparison sentence ("He is 8 years older than Jill") tell me about a comparison. This problem is comparing Joe's age to Jill's age. (*Check off second box under Step 1.*)

Now I am ready for Step 2: Organize the information using the compare diagram. (*Display Compare Problem diagram poster.*)

Compare Problem

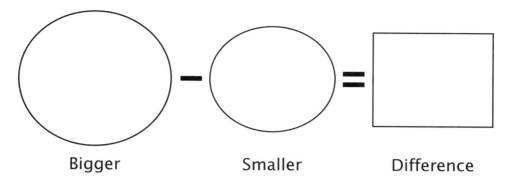

Bigger   Smaller   Difference

(*Point to first box under Step 2.*) To organize the information in a compare problem, we first underline the comparison sentence and circle the two things compared. What is the comparison sentence in this story? How do I know? "He is 8 years <u>older than</u> Jill" (i.e., the second sentence) is the comparison sentence, because the words *older than* (*point to "older than"*) tell about a comparison. Let's underline this sentence as the comparison sentence. (*Pause for students to underline.*) Now we need to circle the two things compared. What are the two things compared in this sentence? (*Hint to students: "He" in the comparison sentence refers to Joe.*)

Students: Joe's and Jill's ages.

Teacher: Circle "Joe" and "Jill" in the comparison sentence. (*Pause for students to circle.*) Remember, the comparison sentence tells us about the difference amount. Do we know the difference between Joe's and Jill's ages from this comparison sentence?

Students: Yes.

Teacher: What is the difference amount?

Students: 8.

Teacher: Great! Circle "8 years" and write it in for the difference amount in the diagram. (*Pause for students to circle.*) Let's check off the first box under Step 2 on the checklist. (*Point to second box under Step 2.*) Next we reread the comparison sentence to find which is the bigger amount (greater quantity) and which is the smaller amount (lesser quantity) and write them in the compare diagram. If Joe is 8 years older than Jill, is the bigger amount Joe's age or Jill's age?

Students: Joe's age.

Teacher: Good. Write "Joe" for the bigger amount and "Jill" for the smaller amount in the compare diagram. (*Pause for students to complete.*) Let's cross out the comparison sentence, because we mapped all the information needed onto our compare diagram. Let's check off the second box under Step 2. (*Point to third box under Step 2.*) Now let's read the story and circle the numbers and labels for the bigger and smaller amounts. The first sentence says, "Joe is 15 years old." Does this sentence tell about the bigger amount or smaller amount? How do you know?

Students: This sentence tells about the bigger amount, because it tells about Joe's age.

Teacher: Good. Underline Joe (bigger amount), circle "15 years," and write in "15 years" for Joe in the diagram. The last sentence in the story says that Jill is 7 years old. What does this sentence tell us? How do you know?

Students: The smaller amount, which is Jill's age.

Teacher: Underline Jill (smaller amount), circle "7 years," and write it in for Jill in the diagram. (*Pause for students to complete.*) We underlined the important information, circled numbers and labels, and wrote the numbers and labels in the compare diagram. Check off the third box under Step 2 on the checklist. Now let's look at the diagram and read what it says.

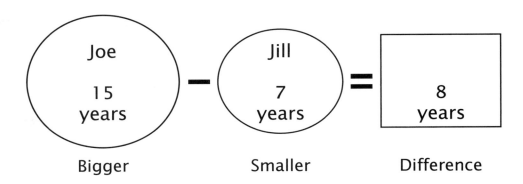

(*Point to relevant parts of diagram as you explain.*) Joe is 15 years old. Jill is 7 years old. The difference in age between Joe and Jill is 8 years. That is, Joe is 8 years older than Jill, or Jill is 8 years younger than Joe. Does this make sense that the difference in age between Joe and Jill is 8 years? How do you know?

Students: This seems to make sense, because if Joe is older than Jill, then the difference of 8 years is correct (i.e., 15 − 7 = 8).

Teacher: That's right. This is a compare problem, because it compares Joe's age to Jill's age (in years).

Compare Story 2

Teacher: (*Display Overhead Modeling page for Compare Story 2. Have students look at student copy of Compare Story 2.*)

Teacher: Touch Story 2. (*Point to checklist.*) What's the first step?

Students: Find the problem type.

Teacher: (*Point to first check box on Compare Story Checklist.*) To find the problem type, I will read the story and retell it in my own words. (*Read story aloud.*)

"Michael has 43 CDs, and Melissa has 70 CDs. Michael has 27 fewer CDs than Melissa."

I read the story. What must I do next?

Students: Retell the story using own words.

Teacher: Yes, I will retell the story in my own words to help me understand it. When I retell, I will ask myself, What do I know in this story? (*Retell the problem.*)

Michael has 27 fewer CDs than Melissa. Michael has 43 CDs, and Melissa has 70 CDs.

I read the story and told it in my own words. (*Check off first box under Step 1 on checklist. Point to second check box under Step 1.*) Now I will ask myself if the story is a compare problem type. Why do you think this is a compare problem? What is this story comparing?

Students: The compare words "fewer than" in the comparison sentence ("Michael has 27 fewer CDs than Melissa") tell me about a comparison. This problem is comparing the number of CDs Michael has to the number of CDs Melissa has.

Teacher: Right! Check off the second box under Step 1. Now I am ready for Step 2: Organize the information using the compare diagram. (*Display Compare Problem diagram poster.*)

Compare Problem

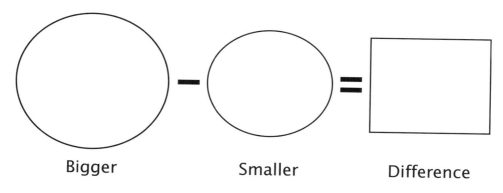

Bigger   Smaller   Difference

(*Point to first box under Step 2.*) To organize the information, we first underline the comparison sentence and circle the two things compared. What is the comparison sentence in this story? How do I know? The second sentence, "Michael has 27 fewer CDs than Melissa," is the comparison sentence, because the words "older than" (*point to "older than"*) tell about a comparison. Let's underline this sentence as the comparison sentence. (*Pause for students to underline.*) Now we need to circle the two things compared. What are the two things compared in this sentence?

Students: The number of CDs that Michael and Melissa have.

Teacher: Circle "Michael" and "Melissa" in the comparison sentence. (*Pause for students to circle.*) What does the comparison sentence tell us? (Is it the difference, smaller, or bigger amount?)

Students: The difference amount.

Teacher: Good! Circle "27 CDs" and write it for the difference amount in the diagram. Let's check off the first box under Step 2 on the checklist. (*Point to second box under Step 2 in checklist.*) Next we reread the comparison sentence to find which is the bigger amount and the smaller amount. If Michael has 27 fewer CDs than Melissa, which is the bigger amount?

Students: The number of CDs Melissa has.

Teacher: Good. Write "Melissa" for the bigger amount and "Michael" for the smaller amount in the compare diagram. (*Pause for students to complete.*) Let's cross out the comparison sentence, because we mapped all the information needed onto our compare diagram. Let's check off the second box under Step 2. (*Point to third box under Step 2.*) Now let's read the story and circle the numbers and labels for the bigger and smaller amounts. The first sentence says, "Michael has 43 CDs, and Melissa has 70 CDs." Does this sentence tell about the bigger amount or smaller amount? How do you know?

Students: This sentence tells about the smaller amount, which is the number of CDs Michael has. It also tells about the bigger amount, which is the number of CDs Melissa has.

Teacher: Underline <u>Michael</u> and <u>Melissa</u> in this sentence. Circle "43 CDs" and write it in for Michael in the diagram. Also, circle "70 CDs" and write it in for Melissa. (*Pause for students to complete.*)

We underlined the important information, circled numbers and labels, and wrote the numbers and labels in the compare diagram. Check off the third box under Step 2 on the checklist. Now let's look at the diagram and read what it says.

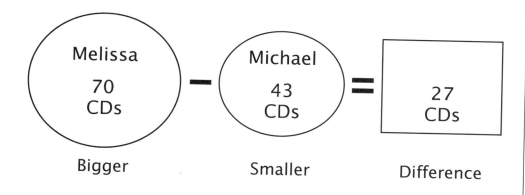

Bigger   Smaller   Difference

Point to relevant parts of the diagram as you explain. Melissa has 70 CDs. Michael has 43 CDs. The difference between the number of CDs that Melissa and Michael have is 27 CDs. That is, Melissa has more CDs than Michael, or Michael has fewer CDs than Melissa. Does this make sense that the difference in the number of CDs between Melissa and Michael is 27? How do you know?

Students: This seems to make sense, because if Melissa has more CDs than Michael, then the difference of 27 CDs is correct (i.e., 70 − 43 = 27).

Teacher: That's right. What is this problem called? Why?

Students: Compare, because it compares the number of CDs Michael has to the number of CDs Melissa has.

Compare Story 3

Teacher: (*Use the script as a guideline for mapping information in Story 3, and facilitate understanding and reasoning by having frequent student–teacher exchanges.*)

(*Display Overhead Modeling page of Compare Story 3. Have students look at student page of Compare Story 3.*)

Touch Story 3. (*Point to the first check box on Compare Story Checklist.*) What's the first step? What do we need to do?

Students: Find the problem type by reading the story and retelling it.

Teacher: Great! (*Read the story aloud or have a student read it.*)
"One week, Anita's horse ate 9 carrots and 7 apples. He ate 2 more carrots than apples that week."
You read the problem. What must you do next?

Students: Retell the story using own words.

Teacher: Yes. Retell the story in your own words to help you understand it. (*Call on students to retell the story. When they retell the story, remind them to tell what they know in the story.*)

Students: One week, Anita's horse ate 2 more carrots than apples. That week, he ate 9 carrots and 7 apples.

Teacher: Check off the first box under Step 1 on the checklist. (*Point to second check box under Step 1.*) What kind of problem is this? How do you know?

Students: Compare, because it is comparing the number of carrots that Anita's horse ate to the number of apples it ate in one week.

Teacher: Correct! (*Check off second box under Step 1 on checklist.*) Now you are ready for Step 2: Organize the information using the compare diagram. To organize the information in a compare story, you first underline the comparison sentence and circle the two things compared. What is the comparison sentence in this story? How do you know?

Students: The comparison sentence is "He ate 2 more carrots than apples that week," because the compare words "more carrots than apples" tell about a comparison. This sentence is comparing the number of carrots eaten to the number of apples eaten.

Teacher: Good. Underline this sentence. (*Pause for students to underline.*) Read this sentence and tell me what two things are compared.

Students: Carrots eaten and apples eaten.

Teacher: Excellent. Circle "carrots" and "apples" in the sentence. (*Pause for students to circle.*) What does the comparison sentence tell us? (Is it the difference, smaller, or bigger amount?)

Students: The difference amount.

Teacher: Do we know the difference amount from this comparison sentence? What is it?

Students: Yes, it is 2.

Teacher: Fantastic! Circle "2" and write "2 more eaten" for the difference amount in the diagram. Check off the first box under Step 2 in the checklist. (*Point to second box under Step 2 of checklist.*) Reread the comparison sentence and find the bigger amount and the smaller amount. Which is the bigger amount? How do you know?

Students: Carrots eaten, because the sentence says that he ate more carrots than apples.

Teacher: Excellent! Circle "carrots" and write "carrots eaten" for the bigger amount in the diagram. Circle "apples" and write "apples eaten" for the smaller amount in the diagram. (*Pause for students to write.*) Let's cross out the comparison sentence, because we mapped all the information needed onto our compare diagram. Check off the second box under Step 2. (*Point to third box under Step 2.*)

Now read the story and circle the numbers and labels for the bigger, smaller, and difference amounts. The first sentence says, "One week,

Anita's horse ate 9 carrots and 7 apples." Does this sentence tell you about the bigger amount or smaller amount? How do you know?

Students: It tells about the bigger and smaller amounts, which are the number of carrots eaten and the number of apples eaten.

Teacher: Excellent. Underline <u>carrots</u> and <u>apples</u>. Circle "9" and write it in for carrots eaten. Also, circle "7" and write it in for apples eaten. *(Pause for students to write.)* We underlined the important information, circled numbers and labels, and wrote the numbers and labels in the compare diagram. Check off the third box under Step 2 on the checklist. Now let's look at the diagram and read what it says.

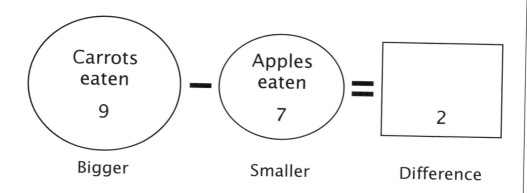

Students: Anita's horse ate 9 carrots and 7 apples. The difference between carrots eaten and apples eaten is 2.

Teacher: That's right. Anita's horse ate 2 more carrots than apples or 2 less apples than carrots. What is this story comparing?

Students: The number of carrots eaten to the number of apples eaten.

Teacher: Does this answer make sense? Why?

Students: Yes, because if Anita's horse ate more carrots than apples, then the difference of 2 is correct (i.e., 9 − 7 = 2).

Teacher: Good. What is this problem called? Why?

Students: Compare, because it compares the number of carrots eaten to the number of apples eaten.

Teacher: *(Pass out Compare Schema Worksheet 1.)* Now I want you to do the next five stories on your own. Remember to use the two steps to organize information in stories using the compare diagram.

(Monitor students as they work. Then check the information in the diagrams. Make sure the diagrams are labeled correctly and completely; see below.)

Compare Schema Story 1: "Kelly scored 21 goals in soccer. She scored 14 fewer goals than Janet. Janet scored 35 goals."

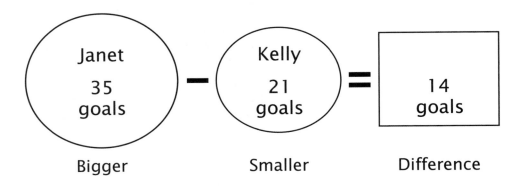

Compare Schema Story 2: "Angie sold 72 magazines for the school fund-raiser. Ed sold 26 fewer magazines than Angie. Ed sold 46 magazines."

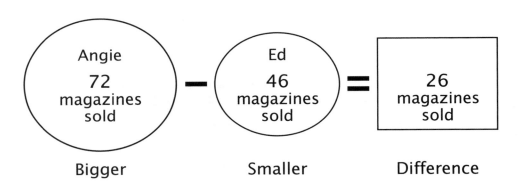

Compare Schema Story 3: "Tina saw 28 movies this summer. She saw 16 more movies than Ruth. Ruth saw 12 movies."

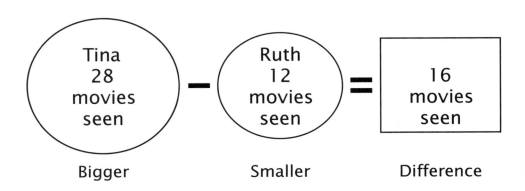

Compare Schema Story 4: "During recess, Katie jumped rope 49 times without stopping. Cindy jumped 81 times without stopping. Katie jumped 32 fewer times than Cindy."

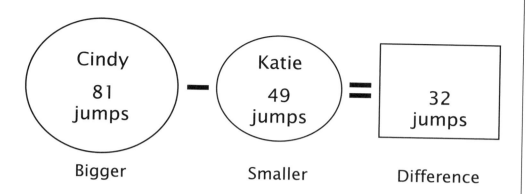

Compare Schema Story 5: "You saw 5 children on the slide and 2 on the swings. There were 3 more children on the slide than on the swings."

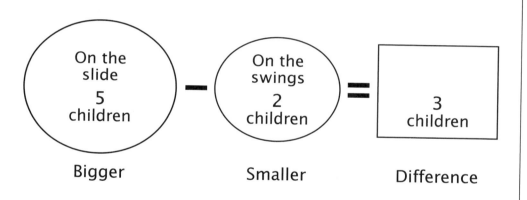

Teacher: You learned to map information in compare story situations onto diagrams. Next you will learn to solve compare problems.

Lesson 12: Problem Solution

Materials Needed

Checklists	Word Problem–Solving Steps (FOPS) poster
	Compare Problem–Solving Checklist (laminated copies for students)
Diagram	Compare Problem diagram poster
Overhead Modeling	Lesson 12: Compare Problems 1, 2, and 3
Reference Guide	Lesson 12: Compare Reference Guide 1
Student Pages	Lesson 12: Compare Problems 1, 2, and 3

Teacher: Today we are going to use compare diagrams like the ones you learned earlier to solve compare word problems. Let's review the compare problem. A compare problem compares two objects, persons, or things using a common unit (e.g., age, height). The comparison sentence in the problem helps us to find the problem type. Words such as *more than, less than, fewer than, younger than, older than, taller than,* and *shorter than,* can help us find the comparison sentence. [*Display Word Problem–Solving Steps (FOPS) poster.*]

Remember the funny word, FOPS. What are the four steps in FOPS?

Students: F—Find the problem type; O—Organize the information in the problem using a diagram; P—Plan to solve the problem; S—Solve the problem.

Compare Problem 1

Teacher: (*Display Overhead Modeling page for Compare Problem 1, and see Compare Reference Guide 1 to set up the problem. Pass out student copies of Compare Problems 1, 2, and 3 and Compare Problem–Solving Checklist. Point to Compare Problem–Solving Checklist.*) We will use this checklist that has the same four steps (FOPS) to help us solve compare problems.

We are ready for Step 1: Find the problem type. (*Point to first check box on Compare Problem–Solving Checklist.*) To find the problem type, I will read the problem and retell it in my own words. Follow along as I read Problem 1. (*Read Compare Problem 1 aloud.*)

"Nathan picked 11 green beans. He picked 7 fewer carrots than green beans. How many carrots did Nathan pick?"

Now I'll retell the problem in my own words to help me understand it. When I retell, I will ask myself, What do I know in this problem and what am I asked to find out? (*Retell the problem.*)

I know that Nathan picked 11 green beans. I also know that he picked 7 *fewer carrots than* green beans. I don't know how many carrots he picked.

I read the problem and told it in my own words. I will check off the first box under Step 1 on the checklist. (*Point to second check box under Step 1.*)

Now I will ask myself if the problem is a compare problem. (*Point to second check box under Step 1.*) Why do you think this is a compare problem? What is this problem comparing?

Students: The compare words *fewer carrots than* in the comparison sentence tell me it is a compare problem. This problem is comparing the number of green beans to the number of carrots that Nathan picked.

Teacher: Let's check off the second box under Step 1. Now I am ready for Step 2: Organize the information in the problem using the compare diagram. (*Display Compare Problem diagram poster.*)

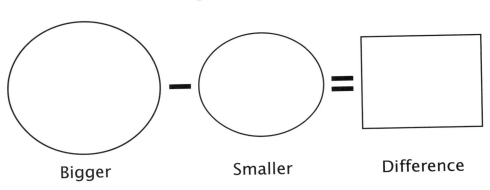

(*Point to first box under Step 2.*) To organize the information in a compare problem, we first underline the comparison sentence or question and circle the two things compared. What is the comparison sentence or question in this story? How do you know?

Students: The second sentence, "He picked 7 fewer carrots than green beans," is the comparison sentence, because the words *fewer than* tell about a comparison.

Teacher: Let's underline this sentence as the comparison sentence. (*Pause for students to underline.*) Now we need to circle the two things compared. What are the two things compared in this sentence?

Students: Green beans and carrots.

Teacher: Circle "green beans" and "carrots" in the comparison sentence. (*Pause for students to circle.*) Remember, the comparison sentence tells us about the difference amount. Do we know the difference between the number of green beans and carrots picked from this comparison sentence?

Students: Yes.

Teacher: What is the difference amount?

Students: 7.

Teacher: Great! Circle "7" and write it in for the difference amount in the diagram. (*Pause for students to circle.*) Let's check off the first box under Step 2 on the checklist. (*Point to second box under Step 2.*) Next we reread the comparison sentence to find which is the bigger amount and which is the smaller amount and write them in the compare diagram. If Nathan picked 7 fewer carrots than green beans, is the bigger amount green beans or carrots?

Students: Green beans.

Teacher: Good. Write in "green beans" for the bigger amount and "carrots" for the smaller amount in the compare diagram. (*Pause for students to complete.*) Let's cross out the comparison sentence, because we mapped all the information needed onto our compare diagram. Let's check off the second box under Step 2. (*Point to third box under Step 2.*) Now let's read the story and circle the numbers and labels for the bigger and smaller amounts. The first sentence says, "Nathan picked 11 green beans." Does this sentence tell about the bigger amount or smaller amount? How do you know?

Students: This sentence tells about the bigger amount, because it tells about the number of green beans picked.

Teacher: Good. Underline green beans (bigger amount), circle "11," and write in "11" for "green beans" in the diagram. The last sentence is the question. (*Point to fourth box under Step 2 on the checklist.*) It asks, "How many carrots did Nathan pick?" This sentence refers to the smaller amount (i.e., number of carrots picked) that we need to solve for. I don't know this amount, so I will write a "?" for it.

Underline carrots (smaller amount), and write a "?" for "carrots in the diagram. (*Pause for students to complete.*) We underlined and circled the important information, circled numbers, and wrote the numbers and labels in the compare diagram. We also wrote a "?" for the smaller amount that we need to solve for. Check off the third and fourth boxes under Step 2 on the checklist. Now let's look at the diagram and read what it says. Nathan picked some carrots and 11 green beans. He picked 7 fewer carrots than green beans. We need to find out the number of green beans that Nathan picked. What must you solve for in this problem? (Is it the bigger, smaller, or difference amount?)

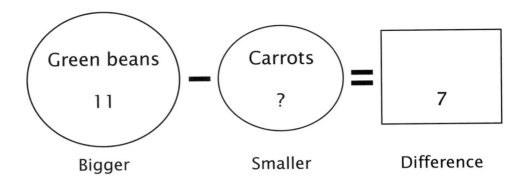

Students: The smaller amount (i.e., the number of carrots picked).

Teacher: Now for Step 3: Plan to solve the problem. (*Point to first box under Step 3 on checklist.*) To plan how to solve the problem, we first decide whether to add or subtract. To do that, I must ask if the total or whole is given. In a compare problem, the total or whole is the bigger amount (or greater quantity). From this diagram, I know the total or whole. If the total or whole is not given, we add; if the total or whole is given, we subtract. The total or whole in this problem is given (*point to diagram*), so do we add or subtract?

Students: Subtract.

Teacher: Good. Let's check off the first box under Step 3 on the checklist. (*Point to second box under Step 3 on checklist.*) Next we write the math sentence. Remember, we have to subtract and when we subtract, we start with the total or whole and subtract the other number. Also, remember to line up the numbers correctly. So the math sentence is 11 − 7 = ?. Let's write the math sentence and check off the second box under Step 3 of the checklist. (*Pause for students to complete.*) Now we are ready for Step 4: Solve the problem. (*Point to first box under Step 4 on checklist.*) What does 11 − 7 =? (*Pause for students to solve the problem.*)

Students: 4.

Teacher: Check off the first box under Step 4 on the checklist. (*Point to second box under Step 4 on checklist.*) Now write "4" for the "?" in the diagram and write the complete answer on the answer line. The complete answer is the number and the label. What is the complete answer to this compare problem?

Students: 4 carrots.

Teacher: Good. Now write "4 carrots" on the answer line. (*Pause for students to write the answer.*) Let's check off the second box under Step 4 on the checklist. (*Point to third box under Step 4 on checklist.*) We are now ready to check the answer. Does 4 green beans seem right? Yes, because Nathan picked fewer carrots (4) than green beans (11). We can also check by adding: 4 + 7 = 11. (*Check off third box under Step 4.*)

(*Model writing the explanation for how Problem 1 was solved here—see Compare Reference Guide 1.*)

Let's review this compare problem. The problem compared the number of green beans to the number of carrots picked. What's this problem called? Why?

Students: Compare, because it compares the number of green beans to the number of carrots picked.

Compare Problem 2

Teacher: (*Display Overhead Modeling page of Compare Problem 2. Have students look at student pages of Compare Problem 2.*)

Touch Compare Problem 2. (*Point to the Compare Problem–Solving Checklist.*) What's the first step?

Students: Find the problem type.

Teacher: Right. (*Point to first check box on checklist.*) To find the problem type, I will read the problem and retell it in my own words. (*Read the problem aloud.*)

"Eric saw a pine tree in the forest. Later Eric saw a maple tree that was 9 feet tall. The maple tree was 5 feet shorter than the pine tree. How tall is the pine tree?"

I read the problem. What must I do next?

Students: Retell the problem using own words and discover the problem type.

Teacher: Yes, I will retell the problem in my own words to help me understand it. When I retell, I will ask myself, What do I know in this problem and what am I asked to find out? (*Retell aloud.*)

I know that Eric saw a maple tree that was 9 feet tall. The maple tree was 5 feet shorter than the pine tree that he saw. I don't know how tall the pine tree is.

I read the problem and told it in my own words. I will check off the first box under Step 1 on the checklist. (*Point to second check box under Step 1.*) Now I will ask myself if the problem is a compare problem. Why do you think this is a compare problem? What is this problem comparing?

Students: The compare words *shorter than* in the comparison sentence tell me it is a compare problem. This problem is comparing the height of the maple tree to the pine tree.

Teacher: Let's check off the second box under Step 1. Now I am ready for Step 2: Organize the information in the problem using the compare diagram. (*Display Compare Problem diagram poster.*)

Compare Problem

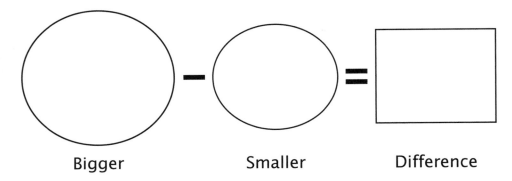

(*Point to first box under Step 2.*) To organize the information in a compare problem, we first underline the comparison sentence or question and circle the two things compared. What is the comparison sentence or question in this story? How do you know?

Students: The third sentence, "The maple tree was 5 feet shorter than the pine tree," is the comparison sentence, because the words *shorter than* tell about a comparison.

Teacher: Let's underline this sentence as the comparison sentence. (*Pause for students to underline.*) Now we need to circle the two things compared. What are the two things compared in this sentence?

Students: The pine tree and the maple tree.

Teacher: Circle "pine tree" and "maple tree" in the comparison sentence. (*Pause for students to circle.*) Remember, the comparison sentence tells us about the difference amount. Do we know the difference between the heights of the maple tree and the pine tree from this comparison sentence?

Students: Yes.

Teacher: What is the difference amount?

Students: 5 feet.

Teacher: Right! Circle "5 feet" and write it for the difference amount in the diagram. (*Pause for students to complete.*) Check off the first box under Step 2. (*Point to second box under Step 2 in checklist.*)

Next we reread the comparison sentence to find which is the bigger amount and which is the smaller amount and write them in the compare diagram. If the maple tree was 5 feet shorter than the pine tree, is the bigger amount the maple tree or the pine tree?

Students: The pine tree.

Teacher: Good. Write in "pine tree" for the bigger amount and "maple tree" for the smaller amount in the compare diagram. (*Pause for students to complete.*) Let's cross out the comparison sentence, because we mapped all the information needed onto our compare diagram. Let's check off the second box under Step 2. (*Point to third box under Step 2.*)

Now let's read the story and circle the numbers and labels for the bigger and smaller amounts. The first sentence says, "Eric saw a pine tree in the forest." This sentence talks about the pine tree, but does not tell us the bigger amount. So we will cross out this sentence. The second sentence says, "Later Eric saw a maple tree that was 9 feet tall." Does this sentence tell about the bigger amount or smaller amount? How do you know?

Students: This sentence tells about the smaller amount, because it tells about the height of the maple tree.

Teacher: Good. Underline maple tree (smaller amount), circle "9 feet," and write in "9 feet" for "maple tree" in the diagram. (*Point to fourth box under Step 2 on checklist.*) The last sentence in the question asks, "How tall is the pine tree?" This sentence refers to the bigger amount (i.e., height of pine tree) that we need to solve for. I don't know this amount, so I will write a "?" for it.

Underline pine tree (bigger amount), and write "? feet" for "pine tree" in the diagram. (*Pause for students to complete.*) We underlined and circled the important information, circled numbers and labels, and wrote the numbers and labels in the compare diagram. We also wrote a "?" for the bigger amount we need to solve for. Check off the third and fourth boxes under Step 2 on the checklist.

Now let's look at the diagram and read what it says. Eric saw a pine tree and a maple tree. The maple tree was 9 feet tall. The maple tree was 5 feet shorter than the pine tree. We need to find out how tall the pine tree is that Eric saw. What must you solve for in this problem? (Is it the bigger, smaller, or difference amount?)

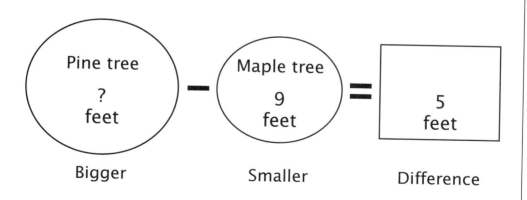

Students: The bigger amount (i.e., the height of the pine tree).

Teacher: Now for Step 3: Plan to solve the problem. (*Point to first box under Step 3 on checklist.*) To plan how to solve the problem, we first decide whether to add or subtract. To do that, I must ask if the total or whole is given. In a compare problem, the total or whole is the bigger amount (or greater quantity). From this diagram, I don't know the total or whole. If the total or whole is not given, we add; if the total or whole is given, we subtract. The total or whole in this problem is given (*point to diagram*), so do we add or subtract?

Students: Add.

Teacher: Good. Let's check off the first box under Step 3 on the checklist. (*Point to second box under Step 3.*) Next we write the math sentence. Remember, we have to add and we have to find the total or whole. Also, remember to line up the numbers correctly. So the math sentence is $9 + 5 = ?$. Let's write the math sentence and check off the second box under Step 3 on the checklist. (*Pause for students to complete.*)

Now we are ready for Step 4: Solve the problem. (*Point to first box under Step 4 on checklist.*) What does $9 + 5 = ?$ (*Pause for students to solve the problem.*)

Students: 14.

Teacher: Check off the first box under Step 4 on the checklist. (*Point to second box under Step 4.*) Now write "14" for the "?" in the diagram and write the complete answer on the answer line. The complete answer is the number and the label. What is the complete answer to this compare problem?

Students: 14 feet.

Teacher: Good. Write "14 feet" on the answer line. (*Pause for students to write the answer.*) Let's check off the second box under Step 4 on the checklist. (*Point to third box under Step 4.*) We are now ready to check the answer. Does 14 feet seem right? Yes, because the pine tree (14 feet) is taller than the maple tree (9 feet). We can also check by subtracting: $14 - 9 = 5$. (*Check off third box under Step 4.*)

(*Guide students to write the explanation for solving the problem on their worksheet; see Compare Reference Guide 1.*) Let's review this compare problem. This problem compared the height of the two trees that Eric saw. What's this problem called? Why?

Students: Compare, because it compares the height of the two trees that Eric saw.

Compare Problem 3

Teacher: (*Use the script as a guideline for solving Compare Problem 3, and facilitate problem solving by having frequent student–teacher exchanges. Display Overhead Modeling page for Compare Problem 3. Have students look at student pages of Compare Problem 3.*)
Touch Compare Problem 3. What's the first step?

Students: Find the problem type.

Teacher: Right. (*Point to first check box on Compare Problem-Solving Checklist.*) To find the problem type, what must you do?

Students: Read the problem and retell it in own words.

Teacher: Good. Read the problem aloud.

Students: "A small bouquet of flowers costs $5.75. It costs $2.75 less for a small bouquet than a large bouquet of flowers. How much does a large bouquet cost?"

Teacher: You read the problem. What must you do next?

Students: Retell it using own words.

Teacher: Yes. Now retell the problem in your own words to help you understand it. (*Call on students to retell the problem. When they retell the problem, remind students to tell what they know in the problem and what they are asked to find out.*)

Students: I know that the small bouquet of flowers costs $5.75. I also know that the small bouquet of flowers costs $2.75 less than the large bouquet of flowers. I don't know how much a large bouquet of flowers costs.

Teacher: Check off the first box under Step 1 on the checklist. (*Point to second check box under Step 1.*) What kind of a problem is this? How do you know?

Students: The compare words *less than* in the comparison sentence tell me it is a compare problem. This problem is comparing the cost of a small bouquet of flowers to a large bouquet of flowers.

Teacher: Check off the second box under Step 1 on the checklist. Now you are ready for Step 2: Organize the information in the problem using the compare diagram. (*Point to first box under Step 2 of checklist.*) To organize the information in a compare problem, what must you do first?

Students: Underline the comparison sentence or question and circle the two things compared.

Teacher: What is the comparison sentence or question in this story? How do you know?

Students: The second sentence, "It costs $2.75 less for a small bouquet than a large bouquet of flowers," is the comparison sentence, because the compare words *less than* tell about a comparison.

Teacher: Underline this comparison sentence. (*Pause for students to underline.*) What must you do next?

Students: Circle the two things compared.

Teacher: What are the two things compared in this problem?

Students: Small bouquet and large bouquet of flowers.

Teacher: Circle "small bouquet" and "large bouquet" in the comparison sentence. (*Pause for students to circle.*) Remember, the comparison sentence tells us about the difference amount. Do you know the difference between the cost of the small and large bouquets from this comparison sentence?

Students: Yes.

Teacher: What is the difference amount?

Students: $2.75.

Teacher: Great! Circle "$2.75" and write it in for the difference amount in the diagram. (*Pause for students to complete.*) Check off the first box under Step 2. (*Point to second box under Step 2 in checklist.*) What do you do next?

Students: Read the comparison sentence to find which is the "bigger" amount and which is the "smaller" amount, and write them in the compare diagram.

Teacher: Which is the bigger amount and which is the smaller amount? How do you know?

Students: The bigger amount is the cost of the large bouquet, because the comparison sentence tells us that the small bouquet costs less than the large bouquet. And the smaller amount is the cost of the small bouquet.

Teacher: Good. Write in "large bouquet" for the bigger amount and "small bouquet" for the smaller amount in the compare diagram. (*Pause for students to complete.*) Now cross out the comparison sentence, because you mapped all the information needed onto the compare diagram. Check off the second box under Step 2. (*Point to third box under Step 2.*) What do you do next?

Students: Read the story and circle the numbers and labels for the bigger and smaller amounts.

Teacher: The first sentence says, "A small bouquet of flowers costs $5.75." Does this sentence tell about the bigger amount or smaller amount? How do you know?

Students: It tells about the smaller amount, because it tells about the cost of the small bouquet.

Teacher: Good. Underline small bouquet (smaller amount), circle "$5.75," and write it in for "small bouquet" in the diagram. The last sentence is the question (*point to fourth box on checklist*) "How much does a large bouquet cost?" This refers to the bigger amount that you need to solve for. You don't know this amount, so write a "?" for it. (*Pause for students to complete.*)

You underlined and circled the important information, circled numbers, and wrote numbers and labels in the compare diagram. You also wrote a "?" for the bigger amount you need to solve for. Check off the third and fourth boxes under Step 2 on the checklist. Now look at the diagram and read what it says.

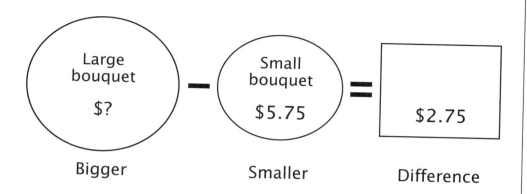

Students: A small bouquet costs $5.75. It is $2.75 less than a large bouquet. We need to find how much a large bouquet costs.

Teacher: What must you solve for in this problem? (Is it the bigger, smaller, or difference amount?)

Students: The bigger amount.

Teacher: Now for Step 3: Plan to solve the problem. To plan how to solve the problem, what must you do? (*Point to first box under Step 3 on checklist.*)

Students: Decide whether to add or subtract.

Teacher: To figure out whether to add or subtract, what should you ask yourself?

Students: If the total or whole is given or not given.

Teacher: In a compare problem, the total or whole is the bigger amount. From this diagram, do you know the total or whole?

Students: No.

Teacher: If the total or whole is not given, do you add or subtract?

Students: Add.

Teacher: Good. Check off the first box under Step 3 on the checklist. What do you do next? (*Point to second box under Step 3.*)

Students: Write the math sentence.

Teacher: Remember, we have to add and we have to find the total or whole. Also, remember to line up the numbers correctly. What is the math sentence?

Students: $5.75 + $2.75 = ?.

Teacher: Write it and check off the second box under Step 3 on the checklist. (*Pause for students to complete.*) Now you are ready for Step 4: Solve the problem. What do you do first?

Students: Solve for $5.75 + $2.75.

Teacher: What does $5.75 + $2.75 =? (*Pause for students to solve the problem.*)

Students: $8.50.

Teacher: Check off the first box under Step 4 on the checklist. What do you do next? (*Point to second box under Step 4.*)

Students: Write the complete answer.

Teacher: The complete answer is the number and the label. What is the complete answer to this compare problem?

Students: $8.50.

Teacher: Good. Write "$8.50" for the "?" in the diagram and write "$8.50" on the answer line. (*Pause for students to write the answer.*) Let's check off the second box under Step 4 on the checklist. (*Point to third box under Step 4.*) What do you do next?

Students: Check the answer.

Teacher: Does $8.50 seem right?

Students: Yes, because the bigger amount ($8.50) is more than the smaller amount ($5.75) or the difference amount ($2.75).

Teacher: We can also check by subtracting: $8.50 − $5.75 = $2.75. (*Check off third box under Step 4.*) Let's review this compare problem. This problem compares the cost of a small bouquet to that of a large bouquet of flowers. What's this problem called? Why?

Students: Compare, because it compares the cost of a small bouquet to that of a large bouquet of flowers.

Teacher: Great job working hard. Tomorrow we will practice more compare problems.

Lesson 13: Problem Solution

Materials Needed

Answer Sheet for Paired-Learning Tasks	Lesson 13: Compare Answer Sheet 1
Checklists	Word Problem–Solving Steps (FOPS) poster
	Compare Problem–Solving Checklist (laminated copies for students)
Diagram	Compare Problem diagram
Overhead Modeling	Lesson 13: Compare Problem 4
Student Pages	Lesson 13: Compare Problem 4
	Lesson 13: Compare Worksheet 1

Compare Problem 4

Teacher: [*Display Word Problem–Solving Steps (FOPS) poster. Ask students to read each step on the poster, which they will use to solve addition and subtraction word problems. Display Overhead Modeling page for Compare Problem 4. Have students look at student pages of Compare Problem 4.*]

Touch Problem 4. (*Point to Compare Problem–Solving Checklist.*) Remember, we will use this checklist that has the same four steps (FOPS) to help us solve compare problems. What's the first step?

Students: Find the problem type.

Teacher: Right. (*Point to first check box on Compare Problem–Solving Checklist.*) To find the problem type, what must you do?

Students: Read the problem and retell it in own words.

Teacher: Good. I will read the problem aloud.
"Use the data from the table to solve the problem."

Cost of a Pack of Tulip Bulbs

Month	Cost
April	$3.85
May	$4.65
June	$5.90

"Stephanie planted a pack of tulip bulbs in April and planted another pack in June. How much more money did Stephanie pay for the tulip bulbs in June than in April?"

I read the problem. What must you do next?

Students: Retell the problem using own words.

Teacher: Yes. Now retell the problem in your own words to help you understand it. (*Call on students to read and retell the problem. When they retell the problem, remind students to tell what they know in the problem and what they are asked to find out.*)

Students: I know how much tulips cost in April, May, and June. I don't know how much more money Stephanie paid for the tulip bulbs in June than in April.

Teacher: Good. Check off the first box under Step 1 on the checklist. (*Point to second check box under Step 1.*) What kind of problem is this? How do you know?

Students: The compare words *more money than* in the comparison question tell me it is a compare problem. This problem is comparing the cost of tulip bulbs in April to the cost of tulip bulbs in June.

Teacher: Check off the second box under Step 1 on the checklist. Now you are ready for Step 2: Organize the information in the problem using the compare diagram. To organize the information, what must you do first?

Students: Underline the comparison sentence or question and circle the two things compared.

Teacher: What is the comparison sentence or question in this story? How do you know?

Students: The question, "How much more money did Stephanie pay for the tulip bulbs in June than in April?" is the comparison question, because the compare words *more than* tell about a comparison.

Teacher: Underline this comparison question. (*Pause for students to underline.*) What must you do next?

Students: Circle the two things compared.

Teacher: What are the two things compared in this problem?

Students: Cost of tulip bulbs in April and June.

Teacher: Circle "tulip bulbs in June" and "in April" in the comparison sentence. (*Pause for students to circle.*) Remember, the comparison sentence tells us about the difference amount. Do you know the difference between the cost of tulip bulbs in June and April from this comparison sentence?

Students: No.

Teacher: Great! Write a "$?" for the difference amount in the diagram. (*Pause for students to complete.*) Check off the first box under Step 2. (*Point to second box under Step 2 in checklist.*) What do you do next?

Students: Read the comparison sentence or question to find which is the "bigger" amount and which is the "smaller" amount and write them in the compare diagram.

Teacher: Which is the bigger amount and which is the smaller amount? How do you know?

Students: The bigger amount is the cost of tulips bulbs in June, because the comparison question tells us that they cost more than tulip bulbs in April. And the smaller amount is the cost of tulip bulbs in April.

Teacher: Good. Write in "tulip bulbs in June" for the bigger amount and "tulip bulbs in April" for the smaller amount in the compare diagram. (*Pause for students to complete.*) Cross out the comparison sentence, because you mapped all the information needed onto the compare diagram. Check off the second box under Step 2. (*Point to third box under Step 2.*) What do you do next?

Students: Read the story and circle the numbers and labels for the bigger and smaller amounts.

Teacher: The first sentence says, "Stephanie planted a pack of tulip bulbs in April and planted another pack in June." Does this sentence tell about the bigger amount or smaller amount?

Students: No.

Teacher: Cross it out. Now read the table to find how much a pack of tulip bulbs costs in April and in June. (*Pause for students to read and figure out.*) What do they cost in April?

Students: $3.85.

Teacher: Good. Underline <u>April</u> (smaller amount), circle "$3.85," and write it in for "tulip bulbs in April" in the diagram. What do they cost in June?

Students: $5.90.

Teacher: Good. Underline <u>June</u> (bigger amount), circle "$5.90," and write it in for "tulip bulbs in June" in the diagram. The table also gives us the information for the cost of a pack of tulip bulbs in May. Do we need this information to solve the problem? Explain.

Students: No, because we are comparing the cost of tulip bulbs in April to the cost in June.

Teacher: Right. Let's cross that out in the table, because we do not need that information to solve our problem. You underlined and circled the important information, circled numbers, and wrote numbers and labels in the compare diagram. You also wrote a "?" for the difference amount we need to solve for. Check off the third and fourth boxes under Step 2 on the checklist. Now look at the diagram and read what it says.

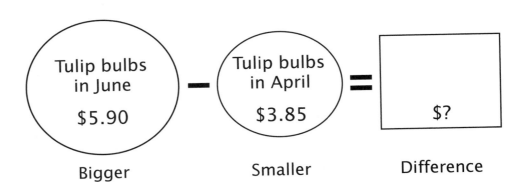

Students: A pack of tulip bulbs costs $3.85 in May and $5.90 in June. We need to find how much more money Stephanie paid for the tulip bulbs in June than in May.

Teacher: What must you solve for in this problem? (Is it the bigger, smaller, or difference amount?)

Students: The difference amount.

Teacher: Now for Step 3: Plan to solve the problem. To plan how to solve the problem, what must you do? (*Point to first box under Step 3 on checklist.*)

Students: Decide whether to add or subtract.

Teacher: To figure out whether to add or subtract, what should you ask yourself?

Students: If the total or whole is given or not given.

Teacher: In a compare problem, the total or whole is the bigger amount. From this diagram, do you know the total or whole?

Students: Yes.

Teacher: If the total or whole is given, do you add or subtract?

Students: Subtract.

Teacher: Good. Check off the first box under Step 3 on the checklist. What do you do next? (*Point to second box under Step 3.*)

Students: Write the math sentence.

Teacher: Remember, you have to subtract, so start with the bigger amount and subtract the other number. Also, remember to line up the numbers correctly. What is the math sentence?

Students: $5.90 - $3.85 = ?$.

Teacher: Write it and check off the second box under Step 3 on the checklist. (*Pause for students to complete.*) Now you are ready for Step 4: Solve the problem. What do you do first?

Students: Solve for $5.90 - $3.85.

Teacher: What does $5.90 - $3.85 =$? (*Pause for students to solve the problem.*)

Students: $2.05.

Teacher: Check off the first box under Step 4 on the checklist. What do you do next? (*Point to second box under Step 4.*)

Students: Write the complete answer.

Teacher: The complete answer is the number and the label. What is the complete answer to this change problem?

Students: Costs $2.05 more in June than in May.

Teacher: Good. Write $2.05 for the "?" in the diagram and write "costs $2.05 more in June than in May" on the answer line. (*Pause for students to write the answer.*) Let's check off the second box under Step 4 on the checklist. (*Point to third box under Step 4.*) What do you do next?

Students: Check the answer.

Teacher: Does $2.05 seem right?

Students: Yes, because the difference amount ($2.05) is the difference between the cost of tulip bulbs in June ($5.90) and in May ($3.85).

Teacher: We can also check by adding the difference and the smaller amount: $2.05 + $3.85 = $5.90. (*Check off third box under Step 4.*) Let's review this compare problem. This problem compares the cost of tulip bulbs in May to the cost of tulip bulbs in June. What's this problem called? Why?

Students: Compare, because it compares the cost of tulip bulbs in May to the cost of tulip bulbs in June.

Teacher: (*Pass out Compare Worksheet 1.*) Now I want you to do Problem 1 on this worksheet with your partner.
   (*Ask students to* think, plan, *and* share *with partners to solve Compare Problem 1 on the worksheet; see Guide to Paired Learning in the Introduction.*)

Compare Worksheet 1, Problem 1: "Today ticket sales were $88 for the circus. This is $34 more than the total sale of yesterday. How much money was collected yesterday?"

   (*Use the four steps to solve the problem. Monitor students as they work. Have students check their answers using the Compare Answer*

Sheet 1. Make sure the diagram is labeled correctly, the math sentence is written and worked out correctly, the written explanation is complete, and the complete answer is written on the answer line; see below.)

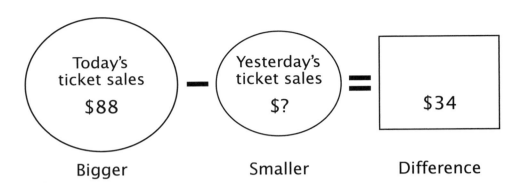

Answer: $54 _____

Teacher: Now I want you to do the next two problems on your own. Remember to use the four steps to solve these problems.

Compare Worksheet 1, Problem 2: "Erin paid $8.99 for a cassette tape on sale. Kate bought the same tape and paid $3.00 more than Erin paid for the tape. How much did Kate pay for the tape?"

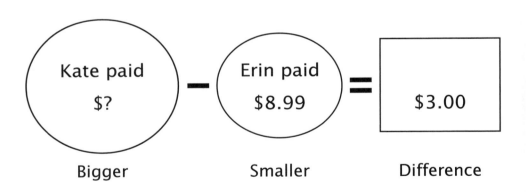

Answer: $11.99 _____

Compare Worksheet 1, Problem 3: "There are 43 more shrubs than trees in the park. Jared counted 98 shrubs in the park. How many trees are there in the park?"

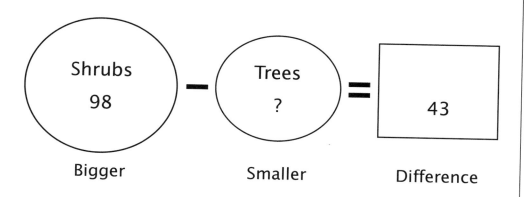

Answer: 55 trees

(*Monitor students as they work. After about 10 minutes, go over the answers. Make sure the diagrams are labeled correctly, the math sentences are worked out correctly, and the complete answers are written on the answer lines.*)

Teacher: Great job working hard. Tomorrow we will practice more compare problems.

Lesson 14: Problem Solution

Materials Needed

Answer Sheet for Paired-Learning Tasks	Lesson 14: Compare Answer Sheet 2
Checklists	Word Problem–Solving Steps (FOPS) poster
	Compare Problem–Solving Checklist (laminated copies for students)
Diagram	Compare Problem diagram poster
Overhead Modeling	Lesson 14: Compare Problems 1 and 2
Student Pages	Lesson 14: Compare Worksheet 2

Teacher: (*Pass out student copies of Compare Worksheet 2. Display Overhead Modeling page of Compare Worksheet 2, Problem 1.*)
Follow along as I read this problem. (*Use guided practice to have students complete Compare Worksheet 2, Problem 1. Read Problem 1 aloud.*)

Compare Worksheet 2, Problem 1. "A redwood tree can grow to be 85 meters tall. A Douglas fir can grow to be 15 meters taller. How tall can the Douglas fir grow?"

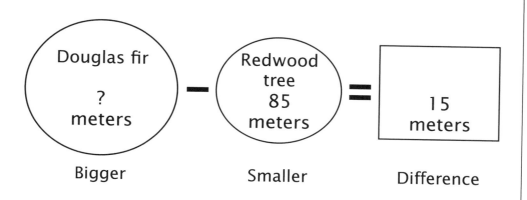

Answer: 100 meters

Teacher: Now I want you to do the next problem with your partner.
(*Ask students to think, plan, and share with partners to solve Compare Worksheet 2, Problem 2; see Guide to Paired Learning in the Introduction.*)

Compare Worksheet 2, Problem 2: "Arthur's brother, Andrew, is 25 inches tall. He is 8 inches shorter than Arthur. How tall is Arthur?"

(*Use the four steps to solve Problem 2. Monitor students as they work. Have students check their answers using Compare Answer Sheet 2. Make sure the diagram is labeled correctly, the math sentence is written and worked out correctly, the written explanation is complete, and the complete answer is written on the answer line; see below.*)

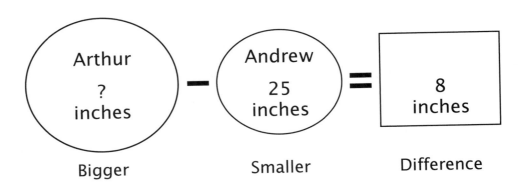

Answer: 33 inches

Teacher: Now I want you to do the next four problems on your own. Remember to use the four steps to solve the problems on this worksheet.

Compare Worksheet 2, Problem 3: "Music Mania sold 56 CDs last week. It sold 29 fewer CDs last week than this week. How many CDs did it sell this week?"

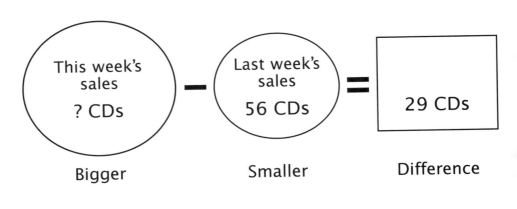

Answer: 85 CDs

Compare Worksheet 2, Problem 4: "India has 3 turtles listed as 'endangered.' Japan has 7 more listed than India. How many turtles are listed as endangered in Japan?"

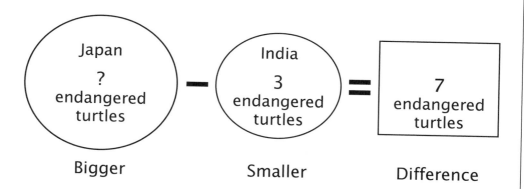

Answer: 10 endangered turtles

Compare Worksheet 2, Problem 5: "Kim's garden has 18 red flowers. There are 5 more yellow flowers than red flowers. How many yellow flowers are in Kim's garden?"

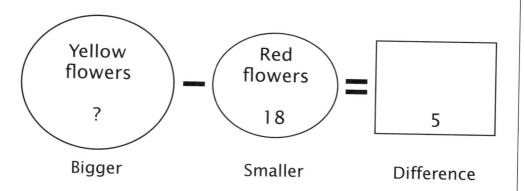

Answer: 23 yellow flowers

"Use the table below to solve Compare Worksheet 2, Problem 6."

People and Their Ages

Name	Age
Larry	55
Ed	27
Joe	39

Compare Worksheet 2, Problem 6: "Larry is 55 years old. Joe is younger than Larry. How many years younger is Joe than Larry?"

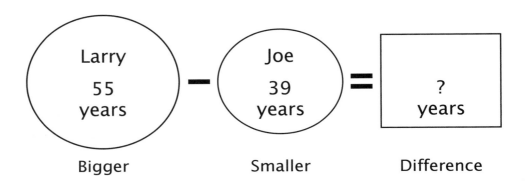

Larry 55 years — Bigger

Joe 39 years — Smaller

? years — Difference

Answer: 16 years

Teacher: (Monitor students as they work. After about 15 to 20 minutes, go over the answers. Make sure the diagrams are labeled correctly, the math sentences are written and worked out correctly, and the complete answers are written on the answer lines.)

Great job working hard. Tomorrow we will solve some more compare problems and get ready to review the three problem types (change, group, and compare).

Lesson 15: Problem Solution

Materials Needed

Checklists
- Word Problem–Solving Steps (FOPS) poster
- Compare Problem–Solving Checklist (laminated copies for students)

Overhead Modeling
Reference Guide
Student Pages
- Lesson 15: Problems 1 and 2
- Lesson 15: Compare Reference Guide 2
- Lesson 15: Compare Worksheet 3

Teacher: (*Pass out student copies of Compare Worksheet 3. Display Overhead Modeling page for Compare Worksheet Problem 1.*)
You learned to solve compare problems using diagrams. Now we will solve the problems on this worksheet using your own diagrams. This worksheet does not have diagrams. Remember to use the four steps (FOPS) to solve problems on the worksheet. (*Note: Discuss how students can generate a diagram that is more efficient than the one they used, and have them practice solving the problems using diagrams they generate. Also, encourage them to use the Compare Problem–Solving Checklist only as needed. Use guided practice to complete Problems 1 and 2 using own diagrams [see below]. Model how to read information in a pictograph for Compare Worksheet Problem 2.*)

Compare Worksheet 3, Problem 1: "A child's movie ticket costs $5.25. A child's ticket costs $2.25 less than an adult's ticket. What is the cost of an adult's ticket?"

Adult's ticket   Child's ticket
$$\underline{\$?} \quad - \quad \underline{\$5.25} \quad = \quad \underline{\$2.25}$$
$$\quad B \qquad\qquad\quad S \qquad\qquad\quad D$$

Answer: $7.50

"Use data from the pictograph to solve Compare Worksheet 3, Problem 2."

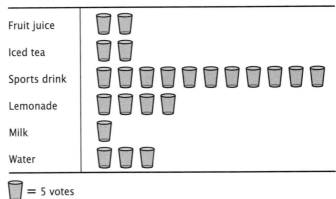

$\square$ = 5 votes

Compare Worksheet 3, Problem 2: "How many more students voted for lemonade than milk?" (*See Compare Reference Guide 2.*)

$$\underset{B}{\underset{\text{students}}{\underset{\text{lemonade}}{\text{Voted for}}\ 20}} - \underset{S}{\underset{\text{students}}{\underset{\text{milk}}{\text{Voted for}}\ 5}} = \underset{D}{\underset{\text{students}}{?}}$$

Answer: 15 more students

Teacher: Now I want you to do the next three problems on your own. (*Have students write the explanation for at least one of the three problems.*) Remember to use the four steps to solve Problems 3 through 5 on the worksheet.

(*Monitor students as they work. After about 15 to 20 minutes, go over the answers. Make sure the diagrams are labeled correctly, the math sentences are worked out correctly, and the complete answers are written on the answer line; see below.*)

Compare Worksheet 3, Problem 3: "Tim is 6 years older than his cousin. If Tim is 21 years old, how old is his cousin?"

$$\underset{B}{\underset{\text{years}}{\underset{\text{Tim}}{}}\ 21} - \underset{S}{\underset{\text{years}}{\underset{\text{Tim's cousin}}{?}}} = \underset{D}{\underset{\text{years}}{6}}$$

Answer: 15 years

Compare Worksheet 3, Problem 4: "Suppose it takes you 70 minutes to walk to school. It takes 45 fewer minutes to ride your bike to school than to walk. How long will it take you to ride your bike to school?"

Walk to school   Ride to school

$$\underline{\underset{B}{\begin{array}{c}70\\ \text{min}\end{array}}} - \underline{\underset{S}{\begin{array}{c}?\\ \text{min}\end{array}}} = \underline{\underset{D}{\begin{array}{c}45\\ \text{min}\end{array}}}$$

Answer: 25 minutes

"Use data from the pictograph to solve Compare Worksheet 3, Problem 5."

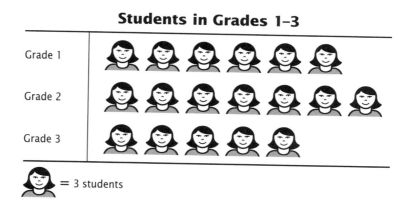

Compare Worksheet 3, Problem 5: "How many more students are in Grade 2 than are in Grade 3?"

Grade 2   Grade 3

$$\underline{\underset{B}{\begin{array}{c}21\\ \text{students}\end{array}}} - \underline{\underset{S}{\begin{array}{c}15\\ \text{students}\end{array}}} = \underline{\underset{D}{\begin{array}{c}?\\ \text{students}\end{array}}}$$

Answer: 6 students

Teacher: You learned to solve compare word problems using your own diagrams. Next you will learn to solve all three types of problems (change, group, and compare) when they are mixed.

Unit 4

One-Step Problem Review
Change, Group, and Compare Problems

Lesson 16: Change, Group, and Compare Review

Materials

Checklists	Change, Group, and Compare Problem–Solving Checklists posters
	Change, Group, and Compare Problem–Solving Checklists (laminated copies for students)
Diagrams	Change, Group, and Compare Problem diagram posters
Overhead Modeling	One-Step Problem Review, Problems 1, 2, and 3
Student Pages	One-Step Problem Review, Problems 1, 2, and 3

Teacher: (*Display posters of Change, Group, and Compare Problem diagrams and checklists in the classroom in a suitable place.*)
Today we are going to review change, group, and compare problems. You will practice these problems to make sure you remember them. Let's review the change problem. What have you learned about a change problem?

Students: A change problem has a beginning, a change, and an ending amount. The beginning, change, and ending all describe the same thing or object.

Teacher: What have you learned about a group problem?

Students: A group problem has two or more small groups or parts that combine to make a large group or whole. Also, the whole is equal to the sum of the parts.

Teacher: What have you learned about a compare problem?

Students: A compare problem compares two objects, persons, or things on a common unit.

Teacher: Great! What are the four steps (FOPS) to solving all addition and subtraction problems?

Students: Find the problem type; Organize information in the problem using diagrams; Plan to solve the problem; and Solve the problem.

Teacher: Good job. (*Point to Change, Group, and Compare Problem diagram posters.*) You learned to use these diagrams to organize information to solve change, group, and compare problems. Later you used your own diagrams to organize information and solve these three different types of problems. Today you will learn to identify each type of problem and solve it using the problem-solving steps you learned earlier. Let's look at Problem 1.

Review Problem 1

Teacher: (*Display Overhead Modeling page for Review Problem 1. Have students look at student copies of Review Problem 1.*) "Use the graph below to solve Review Problem 1."

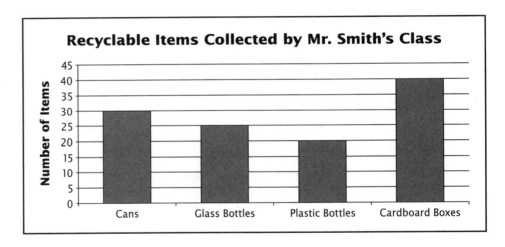

"How many bottles did Mr. Smith's class collect for recycling?"
Touch Review Problem 1. (*Guide students to identify the problem type and solve the problem.*) What's the first step in FOPS?

Students: Find the problem type.

Teacher: Step 1: Find the problem type. To find the problem type, what must you do first?

Students: Read the problem and retell it in own words.

Teacher: Good. (*Read the problem and call on a student to retell the problem.*) Remember, when you retell a problem, you use your own words and tell what you know in the problem and what you are asked to find out.

Students: From this graph, we know the number of glass bottles and plastic bottles recycled. We need to find out how many bottles in all were recycled.

Teacher: Great! You read and retold the problem. Now you need to ask yourself if this is a change, group, or compare problem. What kind of a problem is this? How do you know?

Students: It is a group problem, because it talks about two small groups, glass bottles and plastic bottles, and a large group of all bottles.

Teacher: Now you are ready for Step 2: Organize the information in the problem using the group diagram. To organize the information in your group diagram, what must you do first?

Students: Underline the group names and write them in the diagram.

Teacher: In this problem, how many small groups are there? What are they?

Students: There are two small groups: glass bottles and plastic bottles.

Teacher: Excellent! Write them in your diagram. (*Pause for students to complete.*) What is the large group?

Students: All bottles (or glass and plastic bottles).

Teacher: Good. Write them in your diagram. (*Pause for students to complete.*) Now read the problem and circle the numbers given for the small groups and the large group and write them in your diagram. Also, remember to write a question mark for the unknown information. (*Pause for students to figure out and check their work using the following completely mapped diagram.*)

$$\underset{SG}{\underline{\underset{25}{\text{Glass bottles}}}} + \underset{SG}{\underline{\underset{20}{\text{Plastic bottles}}}} = \underset{LG}{\underline{\underset{?}{\text{All bottles}}}}$$

Teacher: Now look at the diagram and read what it says. (*Pause for students to read.*) What must you solve for in this problem?

Students: The large group amount.

Teacher: Good! You completed Step 2 using FOPS. What is the next step?

Students: Plan to solve the problem.

Teacher: Good. To solve this group problem, do you add or subtract? Why?

Students: We add, because the large group amount (i.e., the total or whole) is not given.

Teacher: Superb! What is the math sentence?

Students: 25 + 20 = ?.

Teacher: Excellent. You completed Step 3 using FOPS. What is Step 4?

Students: Solve the problem.

Teacher: Good. What does 25 + 20 =? (*Pause for students to solve the problem.*)

Students: 45.

Teacher: What is the complete answer to this group problem?

Students: 45 bottles.

Teacher: Good. Write "45 bottles" on the answer line and check your answer. (*Pause for students to write the answer.*) Does 45 seem right? How do you know?

Students: Yes, 45 bottles is the large group and it is more than the two small group amounts (i.e., 25 glass bottles and 20 plastic bottles).

Teacher: Excellent work! Let's solve the next two problems on your worksheet.

Review Problem 2

Teacher: (*Display Overhead Modeling page for Review Problem 2. Have students look at student copies of Review Problem 2.*)
"Ted collected some pictures of butterflies in his scrapbook. This week, he added 25 more pictures. Now he has 90 pictures of butterflies in his scrapbook. How many pictures did he have in the beginning?"
Touch Problem 2. (*Guide students to identify the problem type and solve the problem.*) What's the first step in FOPS?

Students: Find the problem type.

Teacher: Step 1: Find the problem type. To find the problem type, what must you do first?

Students: Read the problem and retell it in own words.

Teacher: Good. (*Read the problem and call on a student to retell the problem.*) Remember, when you retell a problem, you use your own words and tell what you know in the problem and what you are asked to find out.

Students: We know that Ted now has 90 pictures of butterflies in his scrapbook (the ending amount) after he added 25 pictures of butterflies to his scrapbook, which is the change amount. We need to find out the number of pictures of butterflies he began with in his scrapbook (the beginning amount).

Teacher: Great! You read and retold the problem. Now you need to ask yourself if this is a change, group, or compare problem. What kind of problem is this? How do you know?

Students: It's a change problem, because it has a beginning, a change, and an ending. They all describe pictures of butterflies.

Teacher: Now you are ready for Step 2: Organize the information in the problem using the change diagram. To organize the information in your change diagram, what must you do first?

Students: Underline the label that talks about the beginning, change, and ending.

Teacher: What is the label describing these in this problem?

Students: Pictures of butterflies.

Teacher: Good, underline <u>pictures of butterflies</u> and write it in the diagram for the beginning, change, and ending. (*Pause for students to write.*) Now read the problem again; underline the important information (i.e., <u>collected pictures</u>, <u>added more pictures</u>, <u>now has</u>); circle the numbers given for the beginning, change, and ending amounts; and write them in your diagram. Also, remember to write a question mark for the unknown information. (*Pause for students to figure out and check their work using the following completely mapped diagram.*)

$$+\ 25$$
<u>Pictures of butterflies</u>
C

$$?$$
<u>Pictures of butterflies</u>
B

$$90$$
<u>Pictures of butterflies</u>
E

Teacher: Now look at the diagram and read what it says. (*Pause for students to read.*) What must you solve for in this problem?

Students: The beginning amount.

Teacher: Good! You completed Step 2 using FOPS. What is the next step?

Students: Plan to solve the problem.

Teacher: Good. To solve this change problem, do you add or subtract? Why?

Students: We subtract, because the total or whole in this change problem is the ending amount (the change involved an increase), which is given.

Teacher: Superb! What is the math sentence?

Students: 90 − 25 = ?.

Teacher: Excellent. You completed Step 3 using FOPS. What is Step 4?

Students: Solve the problem.

Teacher: Good. What does 90 − 25 =? (*Pause for students to solve the problem.*)

Students: 65.

Teacher: Write "65" for "?" in the diagram. What is the complete answer to this change problem?

Students: 65 pictures of butterflies.

Teacher: Good. Write "65 pictures of butterflies" on the answer line and check your answer. (*Pause for students to write the answer.*) Does 65 seem right? How do you know?

Students: Yes, 65 pictures is less than the ending amount of 90 pictures, which is the total or whole in this problem.

Teacher: We can also check by adding the numbers on the left side of our equation (*point to diagram*): 65 + 25 = 90. What's this problem called? Why?

Students: Change, because it has a beginning, change, and ending amount.

Review Problem 3

Teacher: (*Display Overhead Modeling page for Review Problem 3. Have students look at student copies of Review Problem 3.*)
 "At top speed, a giraffe can run 32 miles an hour. This speed is 3 miles an hour faster than that of an antelope. How many miles an hour does the antelope run?"
 Touch Problem 3. (*Guide students to identify the problem type and solve the problem.*) What's the first step in FOPS?

Students: Find the problem type.

Teacher: Step 1: Find the problem type. To find the problem type, what must you do first?

Students: Read the problem and retell it in own words.

Teacher: Good. (*Read the problem and call on a student to retell the problem.*)

Students: We know that a giraffe can run 32 miles an hour, which is 3 miles an hour faster than an antelope. We don't know how many miles an hour an antelope can run.

Teacher: Great! You read and retold the problem. Now you need to ask yourself if this is a change, group, or compare problem. What kind of problem is this? How do you know?

Students: A compare problem, because the compare words *faster than* in the comparison sentence tell us it is a compare problem. This problem is comparing the miles run in an hour by a giraffe to the miles run in an hour by an antelope.

Teacher: Now you are ready for Step 2: Organize the information in the problem using the change diagram. To organize the information in your compare diagram, what must you do first?

Students: Underline the comparison sentence or question and circle the two things compared.

Teacher: What is the comparison sentence or question in this story? How do you know?

Students: The second sentence, "This speed is 3 miles an hour faster than that of an antelope," is the comparison sentence, because the words *faster than* tell about a comparison.

Teacher: Underline this sentence as the comparison sentence. (*Pause for students to underline.*) What must you do next?

Students: Circle the two things compared in the comparison sentence.

Teacher: What are the two things compared in this problem? (*If students have difficulty, explain that "This speed" in the comparison sentence refers to the miles run in an hour by the giraffe.*)

Students: Miles run in an hour by a giraffe and miles run in an hour by an antelope.

Teacher: Write "giraffe's speed" for "this speed" in the comparison sentence to refer to the giraffe's speed and circle "giraffe" and "antelope." Remember, the comparison sentence also tells us about the difference amount. Do we know the difference between the giraffe's speed and the antelope's speed from this comparison sentence? What is it?

Students: Yes, 3 miles an hour.

Teacher: Super! Circle "3 miles an hour" and write it in for the difference amount. What do you do next?

Students: Read the comparison sentence to find which is the larger amount and which is the smaller amount and write them in the compare diagram.

Teacher: Which is the larger amount and which is the smaller amount? How do you know?

Students: The larger amount is the "giraffe's speed," because the comparison question tells us that the giraffe runs faster than an antelope. The smaller amount is the "antelope's speed."

Teacher: Good. Write in "giraffe's speed" for the larger amount and "antelope's speed" for the smaller amount in the compare diagram. (*Pause for students to complete.*) Cross out the comparison sentence, because you mapped all the information needed onto the compare diagram.

Now read the problem again, circle the numbers given for the larger and smaller amounts, and write them in your diagram. Also, remember to write a question mark for the unknown information. (*Pause for students to figure out and check their work using the following completely mapped diagram.*)

$$\underbrace{\underset{\text{Giraffe's speed}}{\dfrac{32}{\text{miles an hour}}}}_{L} - \underbrace{\underset{\text{Antelope's speed}}{\dfrac{?}{\text{miles an hour}}}}_{S} = \underbrace{\dfrac{3}{\text{miles an hour}}}_{D}$$

Now look at the diagram and read what it says. (*Pause for students to read.*) What must you solve for in this problem?

Students: The small amount.

Teacher: Good! You completed Step 2 using FOPS. What is the next step?

Students: Plan to solve the problem.

Teacher: Good. To solve this compare problem, do you add or subtract? Why?

Students: We subtract, because the larger quantity is the total or whole in the compare problem, which is given.

Teacher: Superb! What is the math sentence?

Students: $32 - 3 = ?$.

Teacher: Excellent. You completed Step 3 using FOPS. What is Step 4?

Students: Solve the problem.

Teacher: Good. What does $32 - 3 = ?$ (*Pause for students to solve the problem.*)

Students: 29.

Teacher: What is the complete answer to this compare problem?

Students: 29 miles an hour.

Teacher: Good. Write 29 miles an hour on the answer line and check your answer. (*Pause for students to write the answer.*) Does 29 seem right? How do you know?

Students: Yes, the antelope's speed of 29 miles an hour is less than the giraffe's speed of 32 miles an hour, which is the larger amount in this problem.

Teacher: Good, now check by subtracting 32 − 29 = 3. Now check off the second box under Step 4 on the checklist. What's this problem called? Why?

Students: Compare, because it compares the speed of the antelope to that of the giraffe.

Teacher: Great job working hard. Tomorrow you will practice solving more change, group, and compare problems.

Lesson 17: Change, Group, and Compare Review

Materials Needed

Checklists
: Change, Group, and Compare Problem–Solving Checklist posters

  Change, Group, and Compare Problem–Solving Checklists (laminated copies for students)

Diagrams
: Change, Group, and Compare Problem diagram posters

Student Pages
: One-Step Problem Review Worksheet 1

Teacher: (*Display posters of Change, Group, and Compare Problem diagrams and checklists in the classroom in a suitable place. Pass out One-Step Problem Review Worksheet 1.*)

(*Review the three problem types—change, group, and compare—and the problem-solving steps for each problem type. Have students recall and distinguish the steps in the problem-solving checklists, especially for Steps 1, Finding the problem type, and 2, Organizing information using a diagram, for the three different problem types. At this time, students should be able to readily recall the problem-solving steps without looking at their checklists. If they have difficulty remembering the steps, have them practice memorizing the steps as homework.*) I want you to do the four problems on this worksheet on your own. Remember to use the four steps to solve these problems.

(*Monitor students as they work. After some time, go over the answers. Make sure the diagrams are labeled correctly, the math sentences are written and worked out correctly, and the complete answers are written on the answer lines; see below.*)

Review Worksheet 1, Problem 1: "Your teacher made some snacks for the class. There were 8 left after the students ate 14 snacks. How many snacks did the teacher make for the class?"

$$\frac{-\ 14}{\text{snacks}}$$
$$\text{C}$$

$$\frac{?}{\text{snacks}} \qquad \frac{8}{\text{snacks}}$$
$$\text{B} \qquad\qquad \text{E}$$

Math Sentence:

$$\begin{array}{r}14\\+8\\\hline 22\text{ snacks}\end{array}$$

Answer: 22 snacks

Review Worksheet 1, Problem 2: "In one week, Samuel read 35 pages. He read 16 fewer pages than Wes. How many pages did Wes read?"

Wes	Samuel	
? pages read	− 35 pages read	= 16 pages read
L	S	D

Math Sentence:

$$\begin{array}{r}35\\+16\\\hline 51\text{ pages read}\end{array}$$

Answer: 51 pages read

Review Worksheet 1, Problem 3: "Ned had 20 crayons. He got some more crayons from his sister. Now he has 32 crayons. How many crayons did he get from his sister?"

+ ?
crayons
C

20
crayons
B

32
crayons
E

136

Math Sentence:

$$\begin{array}{r} 32 \\ -\ 20 \\ \hline 12 \text{ crayons} \end{array}$$

Answer: 12 crayons

"Use the graph below to solve Review Worksheet 1, Problem 4."

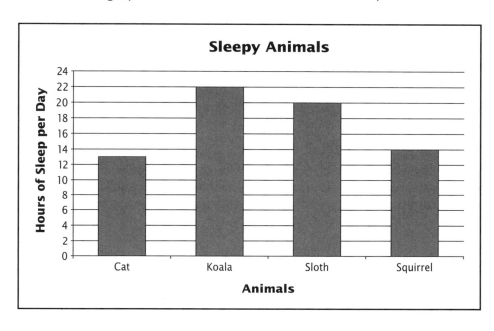

Review Worksheet 1, Problem 4: "How many more hours does the koala sleep than the squirrel?"

Koala		Squirrel			
22 hours	−	14 hours	=	? hours	
L		S		D	

Math Sentence:

$$\begin{array}{r} 22 \\ -\ 14 \\ \hline 8 \text{ hours} \end{array}$$

Answer: 8 hours

Lesson 18: Change, Group, and Compare Review

Materials Needed

Checklists — Change, Group, and Compare Problem–Solving Checklist posters

Diagrams — Change, Group, and Compare Problem diagram posters

Student Pages — Problem Review Worksheet 2

Teacher: (*Display posters of Change, Group, and Compare Problem diagrams and checklists in the classroom in a suitable place. Pass out Problem Review Worksheet 2.*)

Today you will solve five problems on this worksheet on your own. Remember to use the four steps to solve these problems. (*Have students refer to their checklists only as needed.*)

(*Monitor students as they work. After some time, go over the answers. Make sure the diagrams are labeled correctly, the math sentences are written and worked out correctly, and the complete answers are written on the answer lines; see below.*)

Review Worksheet 2, Problem 1: "You have a collection of 12 marbles. If 5 of the marbles in your collection are large, how many marbles are small?"

Large marbles		Small marbles		All marbles
5	+	?	=	12
SG		SG		LG

Math Sentence:

$$\begin{array}{r} 12 \\ -5 \\ \hline 7 \text{ marbles} \end{array}$$

Answer: 7 marbles

Review Worksheet 2, Problem 2: "Olivia has two puzzles. A balloon picture puzzle has 25 pieces. A boat picture puzzle has 5 fewer pieces than the balloon picture puzzle. How many pieces does the boat picture puzzle have?"

Balloon puzzle   Boat puzzle

$$\underline{\text{25 pieces}} - \underline{\text{? pieces}} = \underline{\text{5 pieces}}$$
$$\quad\ \ L \qquad\qquad\quad S \qquad\qquad\quad D$$

Math Sentence:

$$\begin{array}{r} 25 \\ -\ 5 \\ \hline 20 \text{ pieces} \end{array}$$

Answer: 20 pieces

Review Worksheet 2, Problem 3: "David had 25 cents in his wallet. He found 18 cents in his pocket. How much does he have now?"

$$\underline{\text{+ 18 cents}}$$
$$\quad C$$

$$\underline{\text{25 cents}} \qquad\qquad \underline{\text{? cents}}$$
$$\quad B \qquad\qquad\qquad\qquad E$$

Math Sentence:

$$\begin{array}{r} 25 \\ +\ 18 \\ \hline 43 \text{ cents} \end{array}$$

Answer: 43 cents

Review Worksheet 2, Problem 4: "Mr. Lee is 53 years older than his 8-year-old grandson. How old is Mr. Lee?"

Mr. Lee   Mr. Lee's grandson

$$\underline{\underset{L}{\underset{\text{years}}{?}}} - \underline{\underset{S}{\underset{\text{years}}{8}}} = \underline{\underset{D}{\underset{\text{years}}{53}}}$$

Math Sentence:

$$\begin{array}{r} 53 \\ +\ 8 \\ \hline 61 \text{ years} \end{array}$$

Answer: 61 years

Review Worksheet 2, Problem 5: "Karen had 16 of her friends come to her birthday party. 6 of her friends left the party early. How many are still at Karen's birthday party?"

$$\underline{\underset{C}{\underset{\text{friends}}{-\ 6}}}$$

$$\underline{\underset{B}{\underset{\text{friends}}{16}}} \qquad\qquad \underline{\underset{E}{\underset{\text{friends}}{?}}}$$

Math Sentence:

$$\begin{array}{r} 16 \\ -\ 6 \\ \hline 10 \text{ friends} \end{array}$$

Answer: 10 friends

Teacher: Excellent work, everyone! Remember, to be good problem-solvers, use the four steps on your problem-solving checklists. These steps will help you to solve many different addition and subtraction problems. Next you will learn to solve word problems that involve more than one step or problem type.

Unit 5

Two-Step Problems and Mixed Review

Change, Group, and Compare Problems

Lesson 19: Two-Step Change, Group, and Compare Problems

Materials Needed

Checklist	Two-Step Problem–Solving Checklist (laminated copies for students)
Overhead Modeling	Two-Step Problems and Mixed Review, Problems 1, 2, and 3
Student Pages	Two-Step Problems and Mixed Review, Problems 1, 2, and 3

Two-Step Problem 1

Teacher: Today we will learn to solve word problems that involve more than one step or problem type. (*Display Overhead Modeling page of Two-Step Problem 1. Pass out student copies of Two-Step Problems 1, 2, and 3 and Two-Step Problem–Solving Checklist.*)

Let's use the three steps—Ask, Plan, and Solve—on the checklist to solve this problem. Step 1: Ask if the problem is a two-step problem. (*Point to the first check box under Step 1 on the Two-Step Problem–Solving Checklist.*) To do that, I need to read and retell the problem. (*Read the problem aloud.*)

"Use the data in the table below to solve Two-Step Problem 1."

Weight Gained by Bob's Puppy

Month	Weight Gained
January	7 pounds
February	5 pounds
March	3 pounds

"Bob's puppy gained weight from January to March. How many more pounds did the puppy gain in January and February than in March?"

Now I'll retell the problem in my own words to help me understand it. When I retell, I will ask myself, What do I know in this problem and what am I asked to find out? (*Retell the problem.*)

I know that Bob's puppy gained 7 pounds in January, 5 pounds in February, and 3 pounds in March. I don't know how much more weight he gained in January and February together than in March.

I know from reading and retelling that this problem involves more than one step. For me to solve for how much *more* weight Bob's puppy gained in January and February *than* in March, I have to first find the weight gained in January and February.

Let's check off the first box under Step 1. Now I am ready for Step 2: Plan and organize to solve the two-step problem. (*Point to first box under Step 2.*) To solve this problem, I must first find the primary (or main) problem. The primary problem is the problem I must solve for to get the final answer. Sometimes reading the question that the problem asks helps me to identify the primary problem. (*Read the question aloud.*)

"How many more pounds did the puppy gain in January and February than in March?"

This question gives me a clue to figure out the primary (main) problem. Is the primary problem a change, group, or compare problem? How do you know?

Students: A compare problem, because the compare words *more than* in the question refer to a comparison.

Teacher: Right, the primary problem is a compare problem. Let's check off the first box under Step 2 on the checklist. (*Point to second box under Step 2 in checklist.*)

Now I will organize the information in the primary (main) problem using a compare diagram. To organize the information in a compare problem, what do we do first?

Students: Underline the comparison sentence or question and circle the two things compared.

Teacher: Excellent! What is the comparison sentence or question in this story? How do you know?

Students: The question, "How many more pounds did the puppy gain in January and February than in March?" is the comparison statement, because the words *more than* tell about a comparison.

Teacher: Let's underline this as the comparison question. (*Pause for students to underline.*) Next, circle the two things compared. What are the two things compared in this sentence? What are they compared on?

Students: "January and February" and "March." They are compared on the number of pounds gained.

Teacher: Circle "January and February" and "March" in the comparison question. (*Pause for students to circle.*) Also, remember that the comparison sentence tells us about the difference amount. Do we know the difference between the number of pounds gained in January and February and in March from this comparison sentence?

Students: No.

Teacher: Correct, we don't know the difference amount. This is what we are asked to solve for in the problem. That is, we need to solve for the final answer. Let's write a "?" and "pounds" for the difference amount. (*Pause for students to complete.*) What is the next step when you organize information using a compare diagram?

Students: Reread the comparison sentence to find which is the bigger amount and which is the smaller amount, and write them in the compare diagram.

Teacher: Excellent! Which is the bigger amount? Pounds gained in January and February or pounds gained in March? How do you know?

Students: Pounds gained in January and February. I know this is the bigger amount, because from the question I know that he gained more pounds in January and February than in March.

Teacher: Good. Write "January and February" for the bigger amount and "March" for the smaller amount in the compare diagram. You can cross out the comparison question, because you mapped all the information needed onto the compare diagram. (*Pause for students to complete.*) What do you do next to organize information in a compare problem?

Students: Read the problem and circle the numbers and labels for the bigger, smaller, and difference amounts.

Teacher: Now let's read the information in the table to see if we know the bigger amount and the smaller amount. Does the table tell how much weight Bob's puppy gained in January and February?

Students: No.

Teacher: Right! Can you figure out the weight gained in January and February using the information given in the table? Explain.

Students: Yes, because we know the pounds gained in January (7) and the pounds gained in February (5), we can solve for the weight gained in January and February together.

Teacher: Great! So let's write PA (partial answer) for the weight gained in January and February, because we don't know this amount. This is *not* the final answer. However, we can solve for the PA from the information given (i.e., weight gained in January and weight gained in February) in the table. (*Pause for students to complete.*) Does the table tell us how much weight Bob's puppy gained in March?

Students: Yes.

Teacher: How much did he gain in March?

Students: 3 pounds.

Teacher: Underline 3 pounds and write it for the weight gained in March. Now let's look at the diagram and read what it says. (*Bob's puppy gained some pounds in January and February. It gained 3 pounds in March. We need to find out how many more pounds it gained in January and February than in March.*) What must you solve for in this problem to get the final answer? (Is it the bigger, smaller, or difference amount?)

$$\underbrace{\text{PA pounds}}_{B}^{\text{Jan. \& Feb.}} - \underbrace{\text{3 pounds}}_{S}^{\text{March}} = \underbrace{\text{? pounds}}_{D}$$

Students: The difference amount.

Teacher: Excellent! Let's check off the second box under Step 2 on the checklist. (*Point to third box under Step 2 in checklist.*) Before I can solve for the difference amount, I need to first find and solve for the PA (partial answer) or pounds gained in January and February (*point to mapped diagram*). What is the secondary problem that I must solve first to find the final answer to the primary (main) problem? Is the secondary problem to be solved a change, group, or compare? How do you know?

Students: Group, because the two small groups (pounds gained in January and pounds gained in February) combine to make the large group (pounds gained in January and February).

Teacher: Right, the secondary problem is a group problem. Let's check off the third box under Step 2 on the checklist. (*Point to fourth box under Step 2 in checklist.*) Now I will organize the information using a group diagram. To organize the information in a group problem, what do you do first?

Students: Underline the information about the small groups and large group and write the group names in the diagram.

Teacher: Great! In this problem, what are the two small groups?

Students: Pounds gained in January and pounds gained in February.

Teacher: Good, underline January in the table and write it for one of the small groups in the diagram. Underline February and write it in for the other small group in the diagram. What is the large group or whole?

Students: Pounds gained in January and February.

Teacher: Excellent! Write "January and February" for the large group. Because the small groups and the large group all talk about pounds gained, let's write "pounds" in the diagram for all groups. (*Pause for students to write.*)

What must you solve for in this problem? (Is it the small group or large group amount?) (*Remind students about the PA to solve for in the primary problem.*)

Students: The large group.

Teacher: Good. Write "? pounds" for the large group, because we have to solve for it. I know the small groups, pounds gained in January (7) and pounds gained in February (5), from the information in the table. I will underline that information in the table and write it in the diagram.

$$\underline{\underset{SG}{\underset{\text{pounds}}{7}}} + \underline{\underset{SG}{\underset{\text{pounds}}{5}}} = \underline{\underset{LG}{\underset{\text{pounds}}{?}}}$$

Jan.   Feb.   Jan. & Feb.

Let's check off the fourth box under Step 2 on the checklist. Now for Step 3: Solve the problem. (*Point to first box under Step 3 on checklist.*) To solve for the unknown or large group in the secondary problem, do you add or subtract? How do you know?

Students: Add, because the large group quantity is not known.

Teacher: Good. What is the math sentence we must write to solve this group problem?

Students: 7 + 5 = ?.

Teacher: What does 7 + 5 =?

Students: 12.

Teacher: Good. Let's write 12 for the large group or PA in the diagram. How many pounds did the puppy gain altogether in January and February?

Students: 12 pounds.

Teacher: When we solve for the large group amount in this secondary problem, does it give us the final answer?

Students: No.

Teacher: Right. This is the large group amount of 12 pounds, which is the partial answer (PA) to the problem we are asked to solve. Let's write PA for the large group amount and check off the first box under Step 3 on the checklist. (*Point to second box under Step 3.*) Now we will go back to the diagram for the primary (main) or compare problem. Do we know now the number of pounds gained in January and February? What is it?

Students: Yes, 12 pounds.

Teacher: Good. I will cross out "PA" and write "12" for the big amount in the compare diagram. Now I am ready to solve for the unknown in the primary problem. Do you add or subtract to solve for the unknown or the difference amount in the compare problem? How do you know?

Students: Subtract, because the bigger quantity (12 pounds) is known.

Teacher: Great! What is the math sentence?

Students: 12 − 3 = ?.

Teacher: What does 12 − 3 =?

Students: 9.

Teacher: Good. Check off the second box under Step 3 on the checklist. (*Point to third box under Step 3.*) What is the complete or final answer to this problem?

Students: 9 pounds.

Teacher: Excellent. Let's write "9 pounds" on the answer line. (*Pause for students to write the answer.*) Let's check off the third box under Step 3 on the checklist. (*Point to fourth box under Step 3 on the checklist.*) Now I am ready to check the answer. Does 9 seem right? Explain.

Students: Yes, because the difference amount (9 pounds) is the difference between the weight gained in January and February (12) and the weight gained in March (3).

Teacher: Excellent. We can also check by adding the smaller quantity and the difference amount (i.e., 3 + 9 = 12). We are ready to solve the next problem.

Two-Step Problem 2

Teacher: (*Display Overhead Modeling page of Two-Step Problem 2. Have students look at student copies of Two-Step Problem 2.*)
Touch Two-Step Problem 2. What's the first step? (*Point to checklist.*)

Students: Step 1: Ask if the problem is a two-step problem.

Teacher: Right. (*Point to first check box under Step 1 on Two-Step Problem–Solving Checklist.*) To do that, what must you do?

Students: Read and retell the problem.

Teacher: Excellent! I will read the problem. (*Read problem aloud.*)
"Ana goes up 9 steps and then back down 2 steps to pick up a book she dropped. Then she goes up 6 steps. How many steps has she gone up at the end?"
Now I will retell the problem in my own words to help me understand it. (*Retell the problem.*)

I know that Ana first went up 9 steps (beginning amount), then she came down 2 steps (change amount), then she went up 6 steps (more change). I don't know how many steps she climbed (ending amount).

This problem involves more than one step, because to solve for how many steps Ana finally ended up climbing, I have to first solve for the number of steps she ended up with before she climbed 6 more steps.

Let's check off the first box under Step 1, because we read and retold the problem. Now I am ready for Step 2: Plan and organize to solve the two-step problem. (*Point to first box under Step 2.*) To solve this problem, I must first find the primary (main) problem. The primary problem is the problem I must solve for. Sometimes reading the question that the problem asks helps me to identify the primary problem. (*Read the question aloud.*)

In this problem, it is difficult to find the primary problem by only reading the question. I think that retelling the problem in my own words helped me to understand it and identify the problem type. (*Retell the problem if necessary.*)

Ana began by going up 9 steps. This is the beginning amount. Then she went down 2 steps, which describes a change of 2 *less* steps. Next she went up 6 steps. This describes a change of 6 *more* steps. I need to find out the number of steps she ended up climbing.

Also, this problem talks about two change amounts, so it must involve two change problems. The primary and secondary problems are both change problems. Can you tell me why they are change problems?

Students: Because they involve a beginning, a change, and an ending amount. Also, the beginning, change, and ending all talk about steps.

Teacher: Excellent! Let's check off the first and third boxes under Step 2 on the checklist. (*Point out to students that sometimes they do not have to follow the order of the boxes in the checklist to plan and solve the problem.*) Now I will organize the information in the problem using a change diagram. (*Point to second and fourth boxes under Step 2 in checklist.*) To organize the information in a change problem, what do we do first?

Students: Underline the label that talks about the beginning, change, and ending.

Teacher: What is the label describing the beginning, change, and ending in this problem?

Students: Steps.

Teacher: Good, let's underline <u>steps</u> and write it in the diagram for the beginning, change, and ending. (*Pause for students to complete.*) What must you do next to organize information in a change problem?

Students: Underline and circle information about the beginning, change, and ending and write the information in the diagram.

Teacher: What must you solve for in this problem? (Beginning, change, or ending amount?) How do you know?

Students: The ending amount, because the question asks, "How many steps has she gone up?"

Teacher: I will start by writing a question mark for the ending amount, because this is what I must solve for in the problem. Let's read the first sentence, "Ana goes up 9 steps and then back down 2 steps to pick up a book she dropped." Do we know the beginning amount from this sentence? How do you know?

Students: Yes. 9 steps, because when Ana began, she had climbed up 9 steps.

Teacher: Let's underline <u>goes up</u>. Circle "9 steps" and write it for the beginning amount in the diagram. What else do we know from the first sentence?

Students: That Ana came back down 2 steps.

Teacher: Yes, this is telling us about a change of 2 less steps. Let's underline <u>back down</u>. Circle "2" and write "– 2" for the change amount in the diagram. Now let's look at the diagram and read what it says. (Ana began by climbing 9 steps, then she climbed down 2 steps. We need to find out how many steps she climbed at the end.) What must you solve for in this problem? (Is it the beginning, change, or ending amount?)

$$\frac{-\ 2\ \text{steps}}{C}$$

$$\frac{9\ \text{steps}}{B} \qquad \frac{?\ \text{steps}}{E}$$

Students: Ending amount.

Teacher: Do you add or subtract to solve for this ending amount? How do you know?

Students: Subtract, because the beginning amount or total (the change involved a decrease) is known.

Teacher: What is the math sentence? Write and solve for it.

Students: The math sentence is 9 – 2 = ?.

Teacher: What does 9 – 2 =?

Students: 7.

Teacher: The ending amount is 7 steps. Is this ending amount the final answer we are asked to solve for in the problem?

Students: No, because Ana goes up 6 more steps.

Teacher: Right! This ending amount is the partial answer or PA. This is the answer to the secondary problem and not the final answer to the primary (main) problem. Let's write PA for the ending amount and replace the "?" with "7 steps." I organized and solved the secondary problem first in this problem. Let's check off the fourth box under Step 2 and the first box under Step 3, because we organized information in the secondary problem and solved for it.

Now I need to organize information in the primary change problem so I can solve for the final answer. What is the label describing the beginning, change, and ending in the primary problem?

Students: Steps.

Teacher: Good, let's underline steps and write it in the diagram for the beginning, change, and ending. (*Pause for students to complete.*) What must you do next to organize information in a change problem?

Students: Underline and circle information about the beginning, change, and ending and write them in the diagram.

Teacher: What must you solve for in this problem? (Beginning, change, or ending amount?) How do you know?

Students: The ending amount, because the question asks, "How many steps has she gone up at the end?"

Teacher: I will start by writing a question mark for the ending amount, because this is what I must solve for in the problem. Do you know the beginning amount in this problem? How do you know?

Students: Yes, Ana began by climbing 7 steps. This is the PA we got from solving the secondary change problem. This is the beginning amount before Ana climbed 6 more steps.

Teacher: Do we know the change amount? How do you know?

Students: Yes, because the problem tells us that Ana went up 6 steps, which is the change amount.

Teacher: Underline goes up. Circle "6 steps" and write "+ 6" for the change amount in the diagram. Do we know the ending amount?

Students: No, this is the amount we must solve for in this problem.

Teacher: Let's write a question mark for the ending amount or the final answer. Let's look at the diagram for the primary problem and read what it says. (Ana began with 7 steps. Then she climbed 6 *more* steps. We need to find out how many steps she climbed at the end.) What must you solve for in this problem?

$$\frac{+6 \text{ steps}}{C}$$

$$\frac{7 \text{ }\cancel{PA}\text{ steps}}{B} \qquad \frac{?\text{ steps}}{E}$$

Students: Ending amount.

Teacher: Let's check off the second box under Step 2, because we organized the information in the primary problem using a change diagram. Now we are ready to solve for the unknown in the primary problem. (*Point to second box under Step 3 of checklist.*) To solve for the final answer or the ending amount in the primary problem, do you add or subtract? Why?

Students: Add, because the ending amount is the total or whole (the change involved an increase), and it is not known.

Teacher: Great! What is the math sentence?

Students: 7 + 6 = ?.

Teacher: What does 7 + 6 =?

Students: 13.

Teacher: Good, check off the second box under Step 3 on the checklist. (*Point to the third box under Step 3 on checklist.*) What is the complete answer to this problem?

Students: 13 steps.

Teacher: Excellent. Let's write "13 steps" on the answer line. (*Pause for students to write the answer.*) Let's check off the third box under Step 3 on the checklist. (*Point to fourth box under Step 3 on checklist.*) Now I am ready to check the answer. Does 13 seem right? Explain.

Students: Yes, Ana ended up climbing more steps (13) than she began (9).

Teacher: Excellent. We can also check by subtracting 13 − 6 = 7.

Two-Step Problem 3

Teacher: (*Display Overhead Modeling page of Two-Step Problem 3. Have students look at student copies of Two-Step Problem 3. Note: This two-step problem can be solved as change/change or change/group.*)
Touch Two-Step Problem 3. What's the first step? (*Point to checklist.*)

Students: Ask if the problem is a two-step problem.

Teacher: Right. (*Point to first check box under Step 1 on Two-Step Problem–Solving Checklist.*) To figure out if it is a two-step problem, what must you do?

Students: Read and retell the problem.

Teacher: Excellent! I will read the problem. (*Read Two-Step Problem 3 aloud.*)

"Rob is waiting in line to buy snacks. There are 12 people ahead of him. 3 people leave the line without buying anything. 4 people buy their snacks and go to their seats. How many people are ahead of him now?"

(*Call on a student to retell the problem.*)

Students: I know that when Rob was waiting in line to buy snacks, there were 12 people ahead of Rob in the beginning. I know that 3 people then left the line. This is a change of 3 *less* people in line. Also, 4 people left the line after buying their snacks. This is another change of 4 *less* people in line. I need to find the ending amount or the number of people now in line ahead of Rob.

(*Use guided practice to have students complete Two-Step Problem 3 by applying problem-solving procedures similar to those for solving Two-Step Problem 2. Facilitate problem solving by having frequent student–teacher exchanges. See completely mapped information below.*)

Secondary Problem

$$\frac{-3 \text{ people}}{C}$$

$$\frac{12 \text{ people}}{B} \qquad \frac{9 \text{ PA } \cancel{?} \text{ people}}{E}$$

$$12 - 3 = 9$$

Primary Problem

$$\frac{-4}{\text{people}} \over C$$

$$\frac{9 \;\; \cancel{PA}}{\text{people}} \over B \qquad\qquad \frac{?}{\text{people}} \over E$$

$$9 - 4 = 5$$

(Also, discuss an alternative way to solve this problem using change and group problem types. See problem solution below).

Primary Problem

$$\frac{-7 \;\; \cancel{PA}}{\text{people}} \over C$$

$$\frac{12}{\text{people}} \over B \qquad\qquad \frac{5 \;\; \cancel{?}}{\text{people}} \over E$$

$$12 - 7 = 5$$

Secondary Problem

Buy snacks and leave line	Don't buy snacks and leave line	All who leave line
		7 PA
4 people	+ 3 people	= $\cancel{?}$ people
SG	SG	LG

$$4 + 3 = 7$$

Teacher: Great job working hard. Tomorrow we will practice more two-step problems.

Lesson 20: Two-Step Change, Group, and Compare Problems

Materials Needed

Answer Sheets for Paired-Learning Tasks → Two-Step Problem Review Answer Sheet 1

Student Pages → Two-Step Problem Review Worksheet 1

Two-Step Problem Review Worksheet 2

Teacher: Yesterday we learned to solve two-step problems. Now I want you to do the next problem with your partner. (*Pass out Two-Step Problem Review Worksheet 1. Ask students to think, plan, and share with partners to solve Problem 1 [see below] on Two-Step Problem Review Worksheet 1. Also, closely monitor students as they work in pairs.*)

Two-Step Worksheet 1, Problem 1: "Claire went to the school annual book fair. She bought a book that sells for $5, a game for $6, and a mask for $4. How much did she get back from a $20 bill?"

(*Monitor students as they work. Have students check their answers using the Two-Step Problem Review Answer Sheet 1. Make sure the diagrams are labeled correctly, the math sentences are written and worked out correctly, the written explanation is complete, and the complete answer is written on the answer line; see below.*)

Primary Problem

$$-\ \$15$$
$$\cancel{PA}$$
$$\overline{}$$
$$C$$

$$\frac{\$20}{B}$$

$$\$5$$
$$\cancel{?}$$
$$\overline{}$$
$$E$$

$$\$20 - \$15 = \$5$$

Addition and Subtraction

Secondary Problem

Book	Game	Mask	Book, game, & mask
$5	+ $6	+ $4	= $15 PA ~~?~~
SG	SG	SG	LG

$$\$5 + \$6 + \$4 = \$15$$

Answer: $5

Teacher: Now I want you to do the next four problems on your own. *(Pass out Two-Step Worksheet 2. Monitor students as they work. After about 15 minutes, go over the answers. Make sure the diagrams are labeled correctly, the math sentences are written and worked out correctly, and the complete answers are written on the answer lines; see below.)*

Two-Step Worksheet 2, Problem 1: "Tina walks her neighbors' dogs. On Monday, she walks 6 dogs. On Tuesday, she walks 3 more dogs than on Monday. How many dogs does she walk altogether on both days?"

Primary Problem

Monday	Tuesday	Mon. & Tues.
6 dogs	+ 9 ~~PA~~ dogs	= ? dogs
SG	SG	LG

$$6 + 9 = 15$$

Secondary Problem

Tuesday	Monday	Mon. & Tues.
9 PA ~~?~~ dogs	− 6 dogs	= 3 dogs
B	S	D

$$6 + 3 = 9$$

Answer: 15 dogs

Two-Step Worksheet 2, Problem 2: "There are 13 beavers in the lodge and 14 beavers in the pond. If 5 more beavers go into the lodge, how many more beavers are in the lodge than in the pond?"

Primary Problem

In the lodge   In the pond

$\underline{\underset{B}{\overset{\overset{18}{\cancel{PA}}}{\text{beavers}}}} - \underline{\underset{S}{\overset{14}{\text{beavers}}}} = \underline{\underset{D}{\overset{?}{\text{beavers}}}}$

$18 - 14 = 4$

Secondary Problem

$\underline{\underset{C}{\overset{+5}{\text{beavers}}}}$

$\underline{\underset{B}{\overset{13}{\text{beavers}}}} \qquad \underline{\underset{E}{\overset{18 \text{ PA} \;\; \cancel{?}}{\text{beavers}}}}$

$13 + 5 = 18$

Answer: 4 beavers

Two-Step Worksheet 2, Problem 3: "Jane brought in 24 doughnuts for treats during the class play. Joanne brought 12 more doughnuts than Jane. How many doughnuts did the two students bring?"

Primary Problem

Joanne   Jane   Jane & Joanne

$\underline{\underset{SG}{\overset{\overset{36}{\cancel{PA}}}{\text{doughnuts}}}} + \underline{\underset{SG}{\overset{24}{\text{doughnuts}}}} = \underline{\underset{LG}{\overset{?}{\text{doughnuts}}}}$

$36 + 24 = 60$

Addition and Subtraction

Secondary Problem

Joanne   Jane

$$\underset{B}{\underset{\cancel{?}}{\underset{\text{doughnuts}}{36 \text{ PA}}}} - \underset{S}{\underset{\text{doughnuts}}{24}} = \underset{D}{\underset{\text{doughnuts}}{12}}$$

$$24 + 12 = 36$$

Answer: 60 doughnuts

Two-Step Worksheet 2, Problem 4: "Alesha has $15. Arlen has $6 less than Alesha. How much more money does Arlen need to buy a theme park admission ticket that costs $20."

Primary Problem

$$\underset{C}{+\ \$?}$$

$$\underset{B}{\underset{\cancel{\$PA}}{\$9}} \qquad \underset{E}{\$20}$$

$$\$20 - \$9 = \$11$$

Secondary Problem

Alesha   Arlen

$$\underset{B}{\$15} - \underset{S}{\underset{\cancel{?}}{\$9 \text{ PA}}} = \underset{D}{\$6}$$

$$\$15 - \$6 = \$9$$

Answer: $11

Teacher: Great job working hard. Tomorrow we will practice more two-step problems.

Lesson 21: Mixed One-Step and Two-Step Problem Review

Materials Needed

Student Pages   Mixed Review Worksheet 3

Teacher: (*Pass out Mixed Review Worksheet 3.*)
Yesterday you solved two-step problems. Now I want you to do the next few problems on your own. Remember, some of these are one-step problems and other problems involve more than one-step or more than one problem type.

Mixed Review Worksheet 3, Problem 1: "Kelly is learning to play the piano on an electronic keyboard. Her keyboard has 61 keys. Her keyboard has 27 fewer keys than a full-size piano. How many keys does a full-size piano have?"

Full-size piano   Keyboard

$$\underset{B}{\underline{\substack{? \\ \text{keys}}}} - \underset{S}{\underline{\substack{61 \\ \text{keys}}}} = \underset{D}{\underline{\substack{27 \\ \text{keys}}}}$$

$$61 + 27 = 88$$

Answer: 88 keys

Mixed Review Worksheet 3, Problem 2: "Zackery had some party favors to give out at his birthday party. He gave away 35 party favors. Now he has 18 party favors. How many party favors did he have in the beginning?"

$$\underset{C}{\underline{\substack{-35 \\ \text{party favors}}}}$$

$$\underset{B}{\underline{\substack{? \\ \text{party favors}}}} \qquad \underset{E}{\underline{\substack{18 \\ \text{party favors}}}}$$

$$35 + 18 = 53$$

Answer: 53 party favors

Addition and Subtraction

Mixed Review Worksheet 3, Problem 3: "Joy is on an 8-hour car ride. She read a book for 1 hour, saw a DVD movie for 2 hours, and then slept. When she woke up, there were 3 hours to go. How long did she sleep?"

$$\frac{8 \text{ hours}}{B} \quad \frac{-1 \text{ hour}}{C} \quad \frac{7 \text{ PA } \cancel{?} \text{ hours}}{E}$$

$$8 - 1 = 7$$

$$\frac{7 \text{ hours}}{B} \quad \frac{-2 \text{ hours}}{C} \quad \frac{5 \text{ PA } \cancel{?} \text{ hours}}{E}$$

$$7 - 2 = 5$$

$$\frac{5 \cancel{PA} \text{ hours}}{B} \quad \frac{-? \text{ hours}}{C} \quad \frac{3 \text{ hours}}{E}$$

$$5 - 3 = 2$$

Answer: 2 hours

OR

Primary Problem

Read a book	Saw a movie	Slept	Read a book, saw a movie, & slept
$\frac{1 \text{ hour}}{SG}$ +	$\frac{2 \text{ hours}}{SG}$ +	$\frac{? \text{ hours}}{SG}$ =	$\frac{5 \text{ ~~PA~~ hours}}{LG}$

$$5 - (2 + 1) = 2$$

Secondary Problem

$$\frac{5 \text{ PA} - ?\text{ hours}}{C}$$

$$\frac{8 \text{ hours}}{B} \qquad \frac{3 \text{ hours}}{E}$$

$$8 - 3 = 5$$

Answer: 2 hours

"Use the pictograph to answer Mixed Review Worksheet 3, Problem 4."

Students' Favorite Pizza Toppings

Sausage	🍕 🍕
Vegetables	🍕
Extra cheese	🍕 🍕 🍕 🍕
Pepperoni	🍕 🍕 🍕 🍕 🍕 🍕

🍕 = 2 votes

Mixed Review Worksheet 3, Problem 4: "How many students chose vegetables and extra cheese?"

Vegetables		Extra cheese		Vegetables and extra cheese
2 students	+	8 students	=	? students
SG		SG		LG

$$2 + 8 = 10$$

Answer: 10 students

Mixed Review Worksheet 3, Problem 5: "Joshua planned 60 minutes for the Fun House before he has to meet his friends. He waits in line for 25 minutes and spends 20 minutes in the Fun House. How much time does he have left before he has to meet his friends?"

Secondary Problem

	− 25 minutes	
	C	
60 minutes		35 ~~PA~~ minutes
B		E

$$60 − 25 = 35$$

Primary Problem

	− 20 minutes	
	C	
35 ~~PA~~ minutes		? minutes
B		E

$$35 − 20 = 15$$

Answer: 15 minutes

OR

Primary Problem

Time spent		Time left		Total time
45 ~~PA~~ minutes	+	? minutes	=	60 minutes
SG		SG		LG

$$60 - 45 = 15$$

Secondary Problem

Waits in line		In the Fun House		Total time at the Fun House
25 minutes	+	20 minutes	=	45 PA ~~?~~ minutes
SG		SG		LG

$$25 + 20 = 45$$

Answer: 15 minutes

Mixed Review Worksheet 3, Problem 6: "Farmer Moe has 98 animals on his farm. He only has cows and pigs. There are 62 cows on the farm. How many pigs are on the farm?"

Cows		Pigs		Animals on the farm
62	+	?	=	98
SG		SG		LG

$$98 - 62 = 36$$

Answer: 36 pigs

Mixed Review Worksheet 3, Problem 7: "Andrea's dog weighs 65 pounds. It weighs 15 pounds less than Kate's dog. How many pounds does Kate's dog weigh?"

Kate's dog   Andrea's dog

$$\underline{\underset{B}{\overset{?}{\text{pounds}}}} - \underline{\underset{S}{\overset{65}{\text{pounds}}}} = \underline{\underset{D}{\overset{15}{\text{pounds}}}}$$

$$65 + 15 = 80$$

Answer: 80 pounds

Mixed Review Worksheet 3, Problem 8: "You are headed for the moon. It will take 62 hours to get there. If 39 hours have gone by, how many more hours are left to get to the moon?"

$$\underline{\underset{C}{\overset{-\ 39}{\text{hours}}}}$$

$$\underline{\underset{B}{\overset{62}{\text{hours}}}} \qquad \underline{\underset{E}{\overset{?}{\text{hours}}}}$$

$$62 - 39 = 23$$

Answer: 23 hours

"Use the data from the table to solve Mixed Review Worksheet 3, Problems 9 through 11."

Cost of Sports Equipment

Equipment	Cost
Sneakers	$35
Roller blades	$50
Ice skates	$40
Baseball bat	$25
Baseball mitt	$15

Mixed Review Worksheet 3, Problem 9: "Sarah wants a pair of ice skates. She has saved $10. Then her mother gave her $20. How much more money does she need to buy the skates?"

Secondary Problem

$$\begin{array}{r} \$10 \\ +\ \$20 \\ \hline \text{B} \quad \text{C} \end{array}$$

$$\begin{array}{r} \$30 \text{ PA} \\ \cancel{\$?} \\ \hline \text{E} \end{array}$$

$10 + $20 = $30

Primary Problem

$$\begin{array}{r} \$30 \\ \cancel{\$PA} \\ +\ \$? \\ \hline \text{B} \quad \text{C} \end{array}$$

$$\begin{array}{r} \$40 \\ \hline \text{E} \end{array}$$

$40 − $30 = $10

Answer: $10

Mixed Review Worksheet 3, Problem 10: "George wants to play baseball. He saved $60. Then he bought a baseball bat and baseball mitt. How much money does he have now?"

Primary Problem

$$\begin{array}{r} \$60 \\ -\ \$40 \\ -\ \cancel{\$PA} \\ \hline \text{B} \quad \text{C} \end{array}$$

$$\begin{array}{r} \$? \\ \hline \text{E} \end{array}$$

$60 − $40 = $20

Secondary Problem

Baseball bat	Baseball mitt	Baseball bat and mitt
$25	+ $15	= $40 PA ~~$?~~
SG	SG	LG

$$\$25 + \$15 = \$40$$

Answer: $20

Mixed Review Worksheet 3, Problem 11: "Nitasha had $90. She bought sneakers and roller blades. How much money does Nitasha have left?"

Primary Problem

$$
\begin{array}{r}
- \$85 \\
- \cancel{\$PA} \\
\hline
C
\end{array}
$$

$90	$?
B	E

$$\$90 - \$85 = \$5$$

Secondary Problem

Sneakers	Roller blades	Sneakers and roller blades
$35	+ $50	= $85 PA ~~$?~~
SG	SG	LG

$$\$50 + \$35 = \$85$$

Answer: $5

Mixed Review Worksheet 3, Problem 12: "A sea turtle swam 82 miles in the spring. It swam 36 miles more in the spring than in the summer. How many miles did the turtle swim in the summer?"

Spring		Summer		
82 miles	−	? miles	=	36 miles
B		S		D

$$82 - 36 = 46$$

Answer: 46 miles

Teacher: Excellent work, everyone! Remember to use the steps you learned to solve one-step and two-step word problems.

PART II

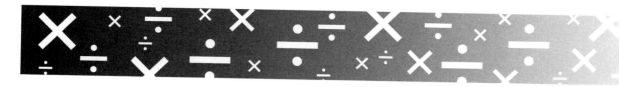

MULTIPLICATION AND DIVISION

Unit 1

Multiplicative Compare Problems

Lesson 1: Problem Schema

Materials Needed

Checklist	Multiplicative Compare (MC) Story Checklist (laminated copies for students)
Diagram	Multiplicative Compare Problem diagram poster
Overhead Modeling	Lesson 1: MC Stories 1, 2, and 3
Student Pages	Lesson 1: MC Stories 1, 2, and 3
	Lesson 1: MC Schema Worksheet 1

Teacher: Today we will learn to identify and organize a type of compare problem that involves multiplication and division rather than addition and subtraction. This type of compare problem is called "multiplicative compare (MC)." An MC problem compares two quantities. Just like the compare problem you learned earlier, an MC problem compares two objects, persons, or things using a common unit (e.g., age, height). However, instead of compare words such as *more than* or *less than* (that involve addition or subtraction), in an MC problem, you will find compare words such as *3 times as many as* or *one half as much as*. In other words, an MC problem tells the quantity of one thing as a multiple (e.g., *2 times as many as*) or part (e.g., *one third*) of the other, and therefore involves the operation of multiplication or division rather than addition or subtraction.

As in a compare problem, the comparison sentence in an MC problem helps us to find the problem type. How do we identify the comparison sentence in an MC problem? Compare words such as *2 times as many as*, or *one third as much as*, or *one half as much as*, can help us find the comparison sentence in the MC problem. For example, "Jim has 5 goldfish. Amy has *4 times as many* goldfish *as* Jim. Amy has 20 goldfish." Which sentence is the comparison sentence and why?

Students: "Amy has 4 times as many goldfish as Jim" is the comparison sentence, because the compare words *4 times as many ... as* tell about a comparison.

Teacher: Yes. Compare words such as *4 times as many ... as* tell about a multiplicative comparison or a multiplicative relation. This is what makes an MC problem different from the addition/subtraction comparison problem you learned earlier. The comparison sentence in MC problems also helps us figure out the two things that are compared in the problem. For example, "Amy has 4 times as many goldfish as Jim." Who are the two people compared in this problem? What are they compared on?

Students: Amy and Jim are compared on the number of goldfish they have.

Teacher: Also, the comparison sentence tells us who is compared to whom. Amy has 4 times as many goldfish as Jim. In this comparison sentence, Amy is compared to Jim on the number of goldfish, so let's label Amy as the *compared* and Jim as the *referent*. Remember, a *referent* is something you compare something else to, or it's also called the benchmark. That is, the referent is the benchmark or standard for comparison. In this compare sentence, "Amy has 4 times as many goldfish as Jim," can you tell me who are the *compared* and the *referent*? How do you know?

Students: Amy is the compared; Jim is the referent, because Amy is compared to Jim on the goldfish they have.

Teacher: Right! Jim is the referent, because Amy is compared to Jim. The comparison sentence also tells us the relation, either a multiple (such as "4 times") or a partial relation (such as "one half"). In this story, the comparison sentence is, "Amy has 4 times as many goldfish as Jim." From this sentence, what is the relation between Amy and Jim with regard to the number of goldfish?

Students: 4 times.

Teacher: Excellent! "4 times" is a multiple relation, because it is a multiple of a whole (one). (*Present a couple more examples orally to help students understand what is the compared and what is the referent when two things are compared. A referent is always something you compare something else to, or the benchmark.*)

(*Display MC Story Checklist.*)

Here are two steps we will use to organize information in an MC story. (*Point to each step on MC Story Checklist and read each one.*) Let's use these two steps to do an example. Look at this story.

MC Story 1

(*Display Overhead Modeling page for MC Story 1. Pass out student copies of MC Stories 1, 2, and 3 and MC Story Checklist. Point to first check box on MC Story Checklist.*)

Now we are ready for Step 1: Find the problem type. To find the problem type, I will read the story and retell it in my own words. (*Read the problem aloud.*)

"Larry has 6 red cars in a box. He has 2 times as many blue cars as red cars. Larry has 12 blue cars."

Now I'll retell the problem in my own words to help me understand it. When I retell, I will ask myself what I know in this story. (*Retell the problem.*)

Larry has both red cars and blue cars. He has 2 times as many blue cars as red cars. That is, the number of blue cars is 2 times as many as red cars. Larry has 12 blue cars and 6 red cars.

I read the story and told it in my own words. Let's check off the first box under Step 1 on the checklist.

Now I will ask myself if the story is an MC problem type. (*Point to second check box under Step 1.*) How do I know it is an MC problem? What is this story comparing? The story is comparing Larry's blue cars to his red cars. The compare words *2 times as many ... as* in the comparison sentence ("He has 2 times as many blue cars as red cars") tell me about a comparison involving a multiple relation (and *not* a "more" or "less" relation as in the compare problem). Therefore, it is a multiplicative compare problem rather than an addition or subtraction compare problem. (*Check off the second box under Step 1.*)

Now I am ready for Step 2: Organize the information in the story using the MC Problem diagram. (*Display diagram.*)

Multiplicative Compare Problem

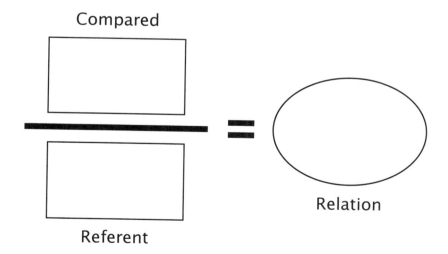

(*Point to first box under Step 2.*) To organize the information in an MC problem, we first underline the comparison sentence, circle the two things compared, and write them in the diagram. What is the comparison sentence in this story? How do I know? "He has 2 times as many blue cars as red cars" (i.e., the second sentence) is the comparison sentence, because the words *2 times as many ... as* (*point to "2 times as many ... as"*) tell about a comparison that involves a multiple relation. Let's underline this sentence as the comparison sentence. (*Pause for students to underline.*) Now we need to circle the two things compared. What are the two things compared in this sentence?

Students: Blue cars and red cars.

Teacher: Circle "blue cars" and "red cars" in the comparison sentence. (*Pause for students to circle.*) Do we know the compared and the referent from this sentence? Remember, the referent is something you compare to or the benchmark.

Students: The number of blue cars is compared to the number of red cars. So blue cars are the compared, and red cars are the referent.

Teacher: Good. Write "blue cars" for the compared and "red cars" for the referent in the MC Problem diagram. (*Pause for students to complete.*)
  Remember, the comparison sentence also tells us about the relation. Do we know the relation between blue cars and red cards from this comparison sentence?

Students: Yes.

Teacher: What is the relation?

Students: 2 times.

Teacher: Great! This tells us that the relation between blue cars and red cars is a multiple relation, because 2 is a multiple of 1 (whole). Circle "2 times" and write it for the relation amount in the diagram. (*Pause for students to circle and write.*) Let's cross out the comparison sentence, because we mapped all the information needed onto our MC diagram. Check off the first box under Step 2 on the checklist.
  (*Point to second box under Step 2.*) Next we reread the story to find the numbers for the compared (i.e., blue cars) and the referent (i.e., red cars) and write them in the MC diagram. The first sentence says, "Larry has 6 red cars in a box." Does this sentence tell about the compare amount or the referent amount?

Students: This sentence tells about the referent amount.

Teacher: Great. "Larry has 6 red cars in a box" tells about the referent amount, because we labeled red cars as the referent in our diagram. (*Refer to the diagram, which indicates "red cars" as the referent.*)
  Now we underline <u>red cars</u>, circle "6," and write "6" for red cars (i.e., the referent) in the diagram. (*Pause for students to complete.*)
  The last sentence in the story says, "Larry has 12 blue cars." Does this tell us the amount for the compared? How do you know?

Students: Yes, because it tells about the number of blue cars, and we labeled "blue cars" as the compared in the diagram.

Teacher: Good. Underline blue cars, circle "12," and write "6" for "blue cars" (i.e., the compared) in the diagram. (*Pause for students to complete.*) We underlined the compared and the referent, circled numbers, and wrote them in the MC diagram. Check off the second box under Step 2 on the checklist. Now let's look at the diagram and read what it says.

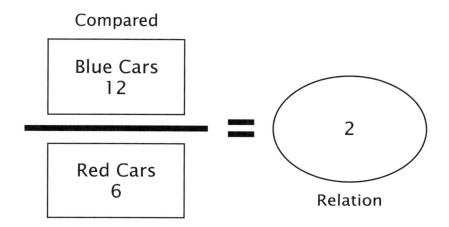

Point to relevant parts of the diagram as you explain. Larry has 2 times as many blue cars as red cars. Larry has 12 blue cars and 6 red cars. 12 is 2 times as many as 6, or 12 to 6 is 2 times (i.e., 12/6 = 2). Does this make sense to you? How do you know?

Students: This seems to make sense, because if the number of blue cars is 12 and the number of red cars is 6, then the number of blue cars is 2 times as many as red cars. 12 is 2 times as much as 6.

Teacher: That's right. This is an MC problem, because it compares the number of Larry's blue cars to his red cars, and the comparison sentence tells about a multiple relation (i.e., 2 times) that involves the operation of multiplication or division.

MC Story 2

Teacher: (*Display Overhead Modeling page of MC Story 2. Have students look at MC Story 2.*)
  Touch Story 2. (*Point to checklist.*) What's the first step?

Students: Find the problem type.

Teacher: (*Point to first check box on MC Story Checklist.*) To find the problem type, I will read the story and retell it in my own words. (*Read story aloud.*)

"Taylor has 125 baseball cards. He has 5 times as many baseball cards as Tony. Tony has 25 baseball cards."

I read the story. What must I do next?

Students: Retell the story using own words.

Teacher: Yes, I will retell the story in my own words to help me understand it. When I retell, I will ask myself, What do I know in this story? (*Retell the problem.*)

Taylor has 125 baseball cards. Taylor has 5 times as many baseball cards as Tony. (*Note. Point out that "He" in the second sentence refers to Taylor.*) Tony has 25 baseball cards.

I read the story and told it in my own words. Check off the first box under Step 1 on the checklist. (*Point to second check box under Step 1.*) Now I will ask myself if the story is an MC problem type. Why do you think this is an MC problem? What is this story comparing?

Students: The compare words *5 times as many ... as* in the comparison sentence ("He has 5 times as many baseball cards as Tony") tell me about a comparison that involves a multiple relation; therefore, it is an MC problem. This problem is comparing the number of baseball cards Taylor has to the number of baseball cards Tony has.

Teacher: Excellent! Check off the second box under Step 1. Now I am ready for Step 2: Organize the information in the story using the MC Problem diagram. (*Display MC Problem diagram.*)

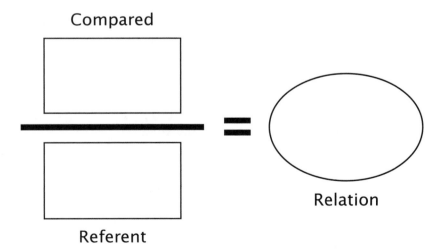

(*Point to first box under Step 2.*) To organize the information, we first underline the comparison sentence, circle the two things compared, and write them in the diagram. What is the comparison sentence in this story? How do I know?

The second sentence, He has 5 times as many baseball cards as Tony, is the comparison sentence, because the words *5 times as many ... as ...* (*point to "5 times as many ... as"*) tell about a comparison that involves a multiple relation (i.e., 5 times). Let's underline this sentence as the comparison sentence. (*Pause for students to underline.*) Now we need to circle the two things compared and write them in the diagram. Who is compared to whom on the number of baseball cards in this sentence?

Students: Taylor is compared to Tony on the number of baseball cards.

Teacher: Circle "he" (i.e., Taylor) and "Tony" in the comparison sentence. (*Pause for students to circle.*) Who are the compared and the referent in the comparison sentence? How do you know?

Students: Taylor is the compared and Tony is the referent, because the compare sentence compares Taylor to Tony on the baseball cards they have.

Teacher: Great. Write "Taylor" in the diagram for the compared and "Tony" for the referent. From this comparison sentence, what is the relation between Taylor and Tony with regard to the number of the baseball cards? Is it a multiple or partial relation?

Students: 5 times, which is a multiple relation.

Teacher: Good! Write "5" in the diagram for the relation, because 5 is a multiple of 1. Let's cross out the comparison sentence, because we mapped all the information needed onto our MC diagram.

Let's check off the first box under Step 2 on the checklist. (*Point to second box under Step 2 on checklist.*) Next we reread the story to find the numbers for the compared (i.e., Taylor) and referent (i.e., Tony). The first sentence says, "Taylor has 125 baseball cards." Does this sentence tell about the compared amount or the referent amount? How do you know?

Students: This sentence tells about the compared amount, because we labeled Taylor as the compared in the diagram.

Teacher: Great! Underline Taylor in this sentence. Circle "125 baseball cards," and write it for Taylor (the compared) in the diagram.

The last sentence in the story says "Tony has 25 baseball cards." Does this sentence tell about the compared amount or the referent amount? How do you know?

Students: This sentence tells about the referent amount, because we labeled Tony as the referent in the diagram.

Teacher: Super! Underline Tony in this sentence. Circle "25 baseball cards," and write it for Tony (the referent) in the diagram. (*Pause for students to complete.*)

We underlined the compared and referent, circled numbers and labels (cards), and wrote them in the diagram. Check off the second box under

Step 2 on the checklist. Now let's look at the diagram and read what it says.

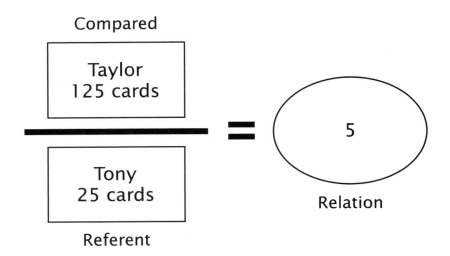

(*Point to relevant parts of diagram as you explain.*) Taylor has 125 baseball cards. Tony has 25 baseball cards. Taylor has 5 times as many baseball cards as Tony. 125 is 5 times as many as 25, or 125 to 25 is 5 times (i.e., 125/25 = 5). Does this make sense that the relation between 125 and 25 is 5 times?

Students: This seems to make sense, because if Taylor has 125 baseball cards and Tony has 25 baseball cards, then comparing 125 to 25 indicates that 125 is 5 times of 25, which is correct.

Teacher: Okay. What is this problem called? Why?

Students: MC, because it compares the number of baseball cards Taylor has to the number Tony has, and the relation is a multiple relation (i.e., 5 times).

Teacher: Correct. A multiple relation (e.g., 5 times) or sometimes a partial relation involves the operation of multiplication or division rather than addition or subtraction.

MC Story 3

Teacher: (*Use the script as a guideline for mapping information in MC Story 3, and facilitate understanding and reasoning by having frequent student–teacher exchanges. Display Overhead Modeling page of MC Story 3. Have students look at MC Story 3.*)
  Touch Story 3. (*Point to first check box on MC Story Checklist.*) What's the first step? What do we need to do?

Students: Find the problem type by reading the story and retelling it.

Teacher: Great! (*Read the story aloud or have a student read it.*)
"Mitch worked 9 hours on the project last week. This week he spent 2/3 as many hours on the project as last week. He spent 6 hours on the project this week."
You read the problem. What must you do next?

Students: Retell the story using own words.

Teacher: Yes. Retell the story in your own words to help you understand it.
(*Note: Because this story involves a fraction, help students understand and retell the story if they have difficulty.*)

Students: Mitch worked 9 hours on the project last week. He worked 6 hours on the project this week. He worked only 2/3 as many hours this week as last week.

Teacher: Great! Check off the first box under Step 1 on the checklist. (*Point to second check box under Step 1.*) What kind of problem is this? How do you know?

Students: MC, because it is comparing the number of hours Mitch worked on the project this week to that of last week.

Teacher: Correct. More important, the story describes a *partial* relation (i.e., 2/3; not "more" or "less" relation) between the number of hours Mitch worked this week and last week. As such, it is an MC rather than an addition or subtraction compare problem. (*Check off second box under Step 1 on checklist.*)
Now you are ready for Step 2: Organize the information in the story using the MC diagram. To organize the information in an MC story, you first underline the comparison sentence, circle the two things compared, and write them in the diagram. What is the comparison sentence in this story? How do you know?

Students: <u>This week he spent 2/3 as many hours on the project as last week</u> is the comparison sentence, because the compare words *2/3 as many ... as* tell about a comparison. This sentence is comparing the number of hours Mitch worked this week to hours worked last week.

Teacher: Good. Underline this sentence. (*Pause for students to underline.*) Read this sentence and tell me what are the two things compared.

Students: Hours worked this week and hours worked last week.

Teacher: Excellent. Circle "this week" and "last week" in the sentence. (*Pause for students to circle.*) What are the compared and the referent in the comparison sentence. How do you know?

Students: "This week" is the compared and "last week" is the referent because the comparison sentence compares the hours Mitch worked this week against the hours he worked last week.

Teacher: Great. Write "this week" in the diagram for the compared and "last week" for the referent. From this comparison sentence, what is the relation between the number of hours Mitch worked this week and the number worked last week? Is it a multiple or partial relation?

Students: 2/3, which is a partial relation.

Teacher: Super! 2/3 is a partial relation, because it is a fraction or a part of one whole. Write "2/3" in the diagram for the relation. Cross out the comparison sentence, because we mapped all the information onto the MC diagram.

Check off the first box under Step 2 in the checklist. (*Point to second box under Step 2 of checklist.*) Now reread the story and find the compared and the referent amounts. What is the number for the compared amount? How do you know?

Students: "6 hours" is the compared amount, because Mitch worked 6 hours this week and we labeled "this week" as the compared in the diagram.

Teacher: Excellent! Underline "this week," circle "6 hours," and write it for "this week" (the compared) in the diagram. What is the number for the referent amount? How do you know?

Students: "9 hours" is the amount for the referent, because Mitch worked 9 hours last week and we labeled last week as the referent in the diagram.

Teacher: Underline <u>last week</u>, circle "9 hours," and write it for the referent amount in the diagram. (*Pause for students to write.*)

We underlined the compared and the referent, circled numbers and labels, and wrote them in the diagram. Let's check off the second box under Step 2. Now look at the diagram and read what it says.

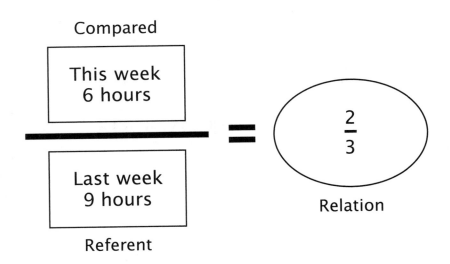

Students: Mitch worked 6 hours on the project this week. He worked 9 hours on the project last week. A comparison of 6 hours to 9 hours indicates that it is 2/3.

Teacher: That's right. Mitch worked less hours on the project this week when compared to last week. 6 is 2/3 of 9 or 6 to 9 is 2 to 3 (6/9 = 2/3). What is this story comparing?

Students: The number of hours Mitch worked on the project this week to that of last week.

Teacher: Does the diagram make sense? Why?

Students: Yes, because if Mitch worked less hours this week, then when we compare the hours he worked this week (a small number, 6) to the hours worked last week (a larger number, 9), or 6 to 9, it is a fraction rather than a multiple. So the partial relation, 2/3, in the comparison sentence seems right.

Teacher: Super! What is this story called? Why?

Students: MC, because the comparison is about a partial relation (2/3), and we know if the comparison involves either a multiple or a partial relation, it is an MC story.

Teacher: (*Pass out Multiplicative Compare Schema Worksheet 1.*) Now I want you to do the next five stories on your own. Remember to use the two steps to organize information in stories using the MC diagram.
(*Monitor students as they work. Then check the information in the diagrams. Make sure the diagrams are labeled correctly and completely; see below.*)

MC Schema Story 1: "Sara completed 24 problems for her math homework. She completed 4 times as many problems as Joe. Joe completed 6 problems."

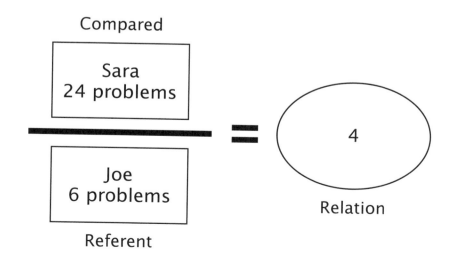

MC Schema Story 2: "Bobby scored 21 goals in soccer. Rickie scored 3 times as many goals as Bobby. Rickie scored 63 goals in soccer."

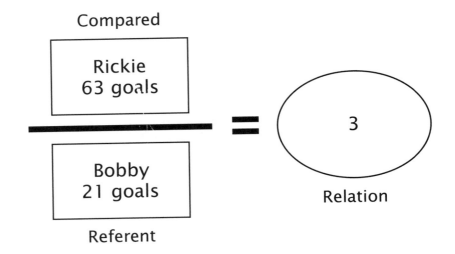

MC Schema Story 3: "Kelly had 32 bottle caps in her collection. She had 2/3 as many caps as Gail. Gail had 48 caps in her collection."

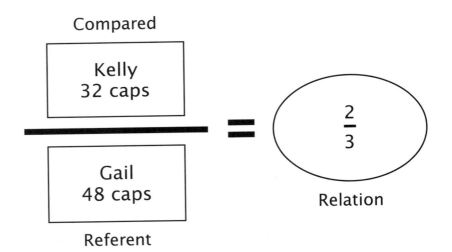

MC Schema Story 4: "The height of a house is 4 meters. The doghouse is only 1/4 as tall as the house. The height of the doghouse is 1 meter."

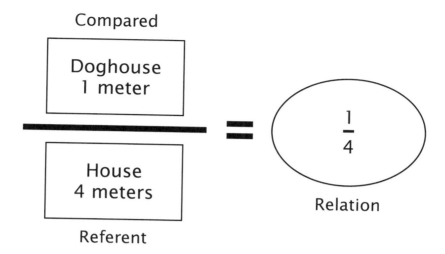

MC Schema Story 5: "Mary has 8 red markers and 4 green markers. She has 2 times as many red markers as green markers."

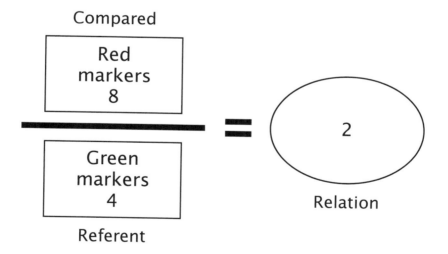

Teacher: You learned to map information in MC story situations onto diagrams. Next you will learn to solve MC problems.

Lesson 2: Problem Solution

Materials Needed

Checklists	Word Problem–Solving Steps (FOPS) poster
	Multiplicative Compare (MC) Problem-Solving Checklist (laminated copies for students)
Diagram	Multiplicative Compare Problem diagram poster
Overhead Modeling	Lesson 2: MC Problems 1, 2, and 3
Reference Guide	Lesson 2: MC Reference Guide 1
Student Pages	Lesson 2: MC Problems 1, 2, and 3

Teacher: Today we are going to use Multiplicative Compare (MC) diagrams like the ones you learned earlier to solve MC word problems. Let's review the MC problem. An MC problem compares two objects, persons, or things using a common unit (e.g., age, height). Specifically, an MC problem tells the quantity of one thing as a multiple (e.g., 3 times as many as) or as a part (e.g., one third) of the other and therefore involves the operation of multiplication or division rather than addition or subtraction.

The comparison sentence in the MC problem helps us to find the problem type. Words such as *3 times as many as, two thirds as much as,* and *one half as much as* help us find the comparison sentence in the MC problem. [*Display Word Problem-Solving Steps (FOPS) poster.*]

Remember the funny word FOPS. What are the four steps in FOPS?

Students: F—Find the problem type; O—Organize the information in the problem using a diagram; P—Plan to solve the problem; S—Solve the problem.

MC Problem 1

(*Display Overhead Modeling page of MC Problem 1. See MC Reference Guide 1 to set up the problem. Pass out student copies of MC Problems 1, 2, and 3 and MC Problem-Solving Checklist.*)

Teacher: (*Point to MC Problem-Solving Checklist.*) We will use this checklist that has the same four steps (FOPS) to help us solve MC problems.

We are ready for Step 1: Find the problem type. (*Point to first check box on MC Problem-Solving Checklist.*) To find the problem type, I will read the problem and retell it in my own words Follow along as I read Problem 1. (*Read MC Problem 1 aloud.*)

"Ray has 4 crayons. Crystal has 5 times as many crayons as Ray. How many crayons does Crystal have?"

Now I'll retell the problem in my own words to help me understand it. When I retell, I will ask myself, What do I know in this problem and what am I asked to find out? (*Retell the problem.*)

I know that Ray has 4 crayons. I also know that Crystal has 5 times as many crayons as Ray. I don't know how many crayons Crystal has.

I read the problem and told it in my own words. I will check off the first box under Step 1 on the checklist.

(*Point to second check box under Step 1.*) Now I will ask myself if the problem is an MC problem. Why do you think this is an MC problem? What is this problem comparing?

Students: The compare words *5 times as many* in the comparison sentence tell me it is an MC problem. This problem is comparing the number of crayons Crystal has to the number of crayons Ray has.

Teacher: Let's check off the second box under Step 1. Now I am ready for Step 2: Organize the information using the MC Problem diagram. (*Display MC Problem diagram poster.*)

Multiplicative Compare Problem

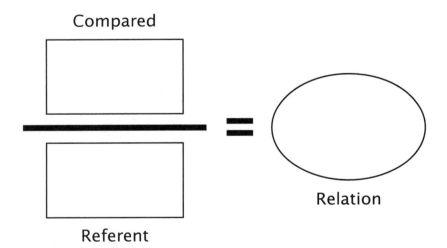

(*Point to first box under Step 2.*) To organize the information in an MC problem, we first underline the comparison sentence or question, circle the two things compared, and write them in the diagram. What is the comparison sentence or question in this story? How do you know?

Students: The second sentence, "Crystal has 5 times as many crayons as Ray," is the comparison sentence, because the words *5 times as many ... as* tell about a comparison involving a multiple relation (i.e., 5 times).

Teacher: Let's underline this sentence as the comparison sentence. (*Pause for students to underline.*) Now we need to circle the two things compared. Who is compared to whom in this sentence? What are they compared on?

Students: Crystal is compared to Ray on the number of crayons.

Teacher: Circle "Crystal" and "Ray" in the comparison sentence. (*Pause for students to circle.*) Who are the compared and the referent in the comparison sentence? How do you know?

Students: "Crystal" is the compared and "Ray" is the referent, because Crystal is compared to Ray on the crayons they have.

Teacher: Great. Write "Crystal" in the diagram for the compared and "Ray" for the referent. From this comparison sentence, what is the relation between Crystal and Ray with regard to the number of crayons? Is it a multiple or a partial relation?

Students: 5 times, which is a multiple relation.

Teacher: Great! Yes, it is a multiple relation, because 5 is a multiple of 1. Circle "5 times" and write it for the relation in the diagram. (*Pause for students to circle.*) Then cross out the comparison sentence, because we mapped all the information onto the MC diagram. Let's check off the first box under Step 2 on the checklist. (*Point to second box under Step 2.*)

Next we reread the problem to find the numbers for the compared (i.e., Crystal) and the referent (i.e., Ray). The first sentence says, "Ray has 4 crayons." Does this sentence tell about the compared amount or the referent amount? How do you know?

Students: This sentence tells about the referent amount, because we labeled Ray as the referent in the diagram.

Teacher: Great! Underline Ray in this sentence. Circle "4 crayons," and write "4 crayons" for "Ray" (the referent) in the diagram.

The last sentence is the question. (*Point to third box under Step 2 on the checklist.*) It asks, "How many crayons does Crystal have?" This sentence is asking for the number of crayons that Crystal has. Is Crystal the referent or the compared?

Students: Crystal is the compared, because we labeled her the compared in the diagram.

Teacher: I don't know this amount, so I will write a "?" for it. Underline Crystal, and write a "?" for "Crystal" (i.e., the compared) in the diagram. (*Pause for students to complete.*) We underlined the compared and referent, circled numbers and labels (crayons), and wrote them in the diagram. We also wrote a "?" for the compared amount that we need to solve for. Check off the second and third boxes under Step 2 on the checklist. Now look at the diagram and read what it says.

Students: Ray has 4 crayons. Crystal has 5 times as many crayons as Ray. We need to find out the number of crayons that Crystal has.

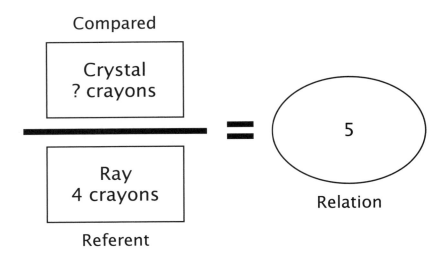

Teacher: Now for Step 3: Plan to solve the problem. (*Point to first box under Step 3 on checklist.*) To plan to solve the problem, all I need to do is translate the information in the diagram into a math equation. That is, ?/4 = 5.

Let's check off the first box under Step 3 on the checklist. Now we are ready for Step 4: Solve the problem. (*Point to first box under Step 4 on checklist.*) We need to solve for the unknown or question mark in the equation.

I can use the multiplication rule for finding equivalent fractions to solve for the unknown or I can directly use cross-multiplication to solve for the unknown. (*Note: Demonstrate the calculation procedure using either the equivalent fraction strategy or the cross-multiplication method. Before students use either strategy, rewrite the equation as follows.*)

$$\frac{?}{4} = \frac{5}{1}$$

Equivalent Fraction Strategy

$$\frac{?}{4} = \frac{5}{1}$$

When you use the equivalent fractions strategy, you multiply or divide the numerator and denominator by the same number to connect one fraction to the other one. To solve for "?", we have to first think "1 times what number gives 4?" The answer is 4 (i.e., 1 × 4 = 4). Then we multiply 5 (the numerator) by the same number to find "?." So 5 × 4 = 20.

Cross-Multiplication Method

$$\frac{?}{4} = \frac{5}{1}$$

By applying the cross-multiplication method, the equation is:

? × 1 = 4 × 5

? = 20

Check off the first box under Step 4 on the checklist. (*Point to second box under Step 4.*) Now write 20 for the "?" in the diagram and write the complete answer on the answer line. The complete answer is the number and the label. What is the complete answer to this MC problem?

Students: 20 crayons.

Teacher: Good. I'll write 20 crayons on the answer line. (*Pause for students to write the answer.*) Let's check off the second box under Step 4 on the checklist. (*Point to third box under Step 4 on the checklist.*) We are now ready to check the answer. Does "20 crayons" seem right?

Yes, because Crystal (20) has 5 times as many crayons as Ray (4). We can also check by redoing the cross-multiplication with all the numbers. That is, 20 × 1 = 4 × 5, or 20 = 20. So my calculation is correct. (*Check off third box under Step 4. Model writing the explanation for how the problem was solved; see MC Reference Guide 1.*)

Let's review this MC problem. This problem compared Crystal to Ray on the number of crayons. What's this problem called? Why?

Students: MC, because the comparison sentence tells about a multiple relation (i.e., 5 times) between the compared and the referent.

MC Problem 2

Teacher: (*Display Overhead Modeling page of MC Problem 2. Have students look at MC Problem 2.*)

Touch Problem 2. (*Point to the MC Problem-Solving Checklist.*) What's the first step?

Students: Find the problem type.

Teacher: Right. (*Point to first check box on checklist.*) To find the problem type, I will read the problem and retell it in my own words. (*Read the problem aloud.*)

"Both Johnny and Greg play Little League baseball. Greg hit 60 home runs. Greg hit 3 times as many home runs as Johnny. How many home runs did Johnny hit during the season?"

I read the problem. What must I do next?

Students: Retell the problem using own words and discover the problem type.

Teacher: Yes, I will retell the problem in my own words to help me understand it. When I retell, I will ask myself, What do I know in this problem and what am I asked to find out? (*Retell aloud.*)

This story is about baseball. It is comparing Greg to Johnny on the number of home runs they hit. I know Greg hit 60 home runs. I also know Greg hit 3 times as many home runs as Johnny. I do not know how many home runs Johnny hit, and that is what I am asked to solve for.

I read the problem and told it in my own words. I will check off the first box under Step 1 on the checklist. (*Point to second check box under Step 1.*)

Now I will ask myself if the problem is an MC problem. Why do you think this is an MC problem? What is this problem comparing?

Students: The compare words *3 times as many ... as* in the comparison sentence tell me it is an MC problem, because it tells about a multiple relation (3 times). This problem is comparing Greg to Johnny on the number of home runs they hit.

Teacher: Let's check off the second box under Step 1. Now I am ready for Step 2: Organize the information in the problem using the MC diagram. (*Display MC Problem diagram poster.*)

Multiplicative Compare Problem

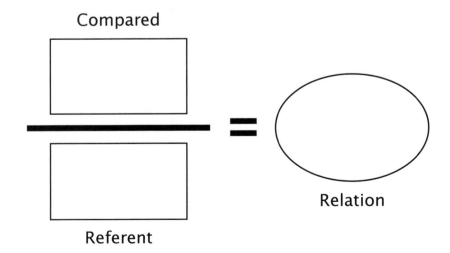

Teacher: *(Point to first box under Step 2.)* To organize the information in an MC problem, we first underline the comparison sentence or question, circle the two things compared, and write them in the diagram. What is the comparison sentence or question in this story? How do you know?

Students: The third sentence, "Greg hit 3 times as many home runs as Johnny," is the comparison sentence, because the words *3 times as many ... as* tell about a comparison involving a multiple relation.

Teacher: Let's underline this sentence as the comparison sentence. *(Pause for students to underline.)* Now we need to circle the two things compared. Who is compared to whom in this sentence? What are they compared on?

Students: Greg is compared to Johnny on the number of home runs hit.

Teacher: Circle "Greg" and "Johnny" in the comparison sentence. *(Pause for students to circle.)* Who are the compared and the referent in the comparison sentence? How do you know?

Students: "Greg" is the compared and "Johnny" is the referent, because Greg is compared to Johnny on the number of home runs hit.

Teacher: Great. Write "Greg" in the diagram for the compared and "Johnny" for the referent. From this comparison sentence, what is the relation between Greg and Johnny with regard to the number of home runs hit? Is it a multiple or partial relation?

Students: 3 times, which is a multiple relation.

Teacher: Great! Yes, it is a multiple relation, because 3 is a multiple of 1. Circle "3 times" and write it for the relation in the diagram. *(Pause for students to complete.)* Let's check off the first box under Step 2 on the checklist. *(Point to second box under Step 2.)* Next we reread the problem to find the numbers for the compared (i.e., Greg) and the referent (i.e., Johnny). The first sentence says, "Both Johnny and Greg play Little League baseball." Does this sentence tell about the compared or referent amount?

Students: No.

Teacher: Let's cross out this sentence, because we don't need this information to solve the problem. The second sentence says, "Greg hit 60 home runs." Does this sentence tell about the compared amount or the referent amount? How do you know?

Students: This sentence tells about the compared amount, because we labeled Greg as the compared in the diagram.

Teacher: Great! Underline Greg in this sentence. Circle "60 home runs" and write it for Greg (the compared) in the diagram.

The last sentence is the question. *(Point to third box under Step 2 on checklist.)* It asks, "How many home runs did Johnny hit during the season?" This sentence is asking for the number of home runs hit by Johnny. Is Johnny the referent or the compared?

Students: Johnny is the referent, as we labeled in the diagram.

Teacher: I don't know this amount, so I will write a "?" for it. Underline Johnny, and write a "?" and the label (home runs) for "Johnny" (i.e., the referent) in the diagram. (*Pause for students to complete.*)

We underlined the compared and the referent, circled numbers and labels (home runs), and wrote them in the diagram. Because the amount for the referent is not given, we wrote "? home runs." This referent amount is what we need to solve for. Check off the second and third boxes under Step 2 on the checklist. (*Pause for students to complete.*) Now look at the diagram and read what it says.

Students: Greg hit 60 home runs. Greg hit 3 times as many home runs as Johnny. We need to find out the number of home runs that Johnny hit.

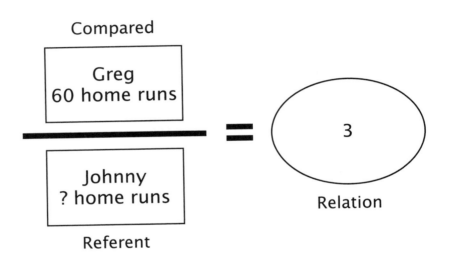

Teacher: Yes. Now for Step 3: Plan to solve the problem. (*Point to first box under Step 3 on the checklist.*) To plan to solve the problem, we need to translate the information in the diagram into a math equation:

$$\frac{60}{?} = 3$$

Let's check off the box under Step 3 on the checklist. Now we are ready for Step 4: Solve the problem. (*Point to first box under Step 4 on checklist.*) We need to solve for the unknown or question mark in the equation.

Let's use the cross-multiplication method to solve for it. Before you use cross-multiplication, how do you rewrite the equation?

Students: $\frac{60}{?} = \frac{3}{1}$

Teacher: Super! Now, we can solve for the "?" by applying cross-multiplication:

$$? \times 3 = 60 \times 1$$
$$? = 60 \div 3$$

Teacher: What is 60 ÷ 3?

Students: 20.

Teacher: Great job! Check off the first box under Step 4 on the checklist. (*Point to second box under Step 4 on checklist.*) Now write 20 for the "?" in the diagram and write the complete answer on the answer line. The complete answer is the number and the label. What is the complete answer to this MC problem?

Students: 20 home runs.

Teacher: Good. I'll write 20 home runs on the answer line. (*Pause for students to write the answer.*) Let's check off the second box under Step 4 on the checklist. (*Point to third box under Step 4 on checklist.*) We are now ready to check the answer. Does "20 home runs" seem right?

Students: Yes, because Greg (60) hit 3 times as many home runs as Johnny (20).

Teacher: We can also check by redoing the cross-multiplication with all the numbers. That is, 60 × 1 = 20 × 3, or 60 = 60. So our calculation is correct. (*Check off third box under Step 4. Guide students to write the explanation for solving the problem on their worksheet; see MC Reference Guide 1.*)

Let's review this MC problem. This problem compared Greg to Johnny on the number of home runs hit. What's this problem called? Why?

Students: MC, because the comparison sentence tells a multiple relation (i.e., 3 times) between the compared and the referent.

MC Problem 3

Teacher: (*Use the script as a guideline for solving MC Problem 3, and facilitate problem solving by having frequent student–teacher exchanges. Display Overhead Modeling page for MC Problem 3. Have students look at MC Problem 3.*)

Touch Problem 3. What's the first step?

Students: Find the problem type.

Teacher: Right. (*Point to first check box on MC Problem-Solving checklist.*) To find the problem type, what must you do?

Students: Read the problem and retell it in your own words.

Teacher: Good. Read the problem aloud.

Students: "Michael sold 18 kites on Saturday. He sold half as many kites on Saturday as on Friday. How many did he sell on Friday?"

Teacher: You read the problem. What must you do next?

Students: Retell it using own words.

Teacher: Yes. Now retell the problem in your own words to help you understand it. (*Call on students to retell the problem. When they retell the problem, remind students to tell what they know in the problem and what they are asked to find out.*)

Students: This problem is about selling kites. I know Michael sold 18 kites Saturday. I also know he sold half as many kites on Saturday as on Friday. I do not know how many he sold on Friday.

Teacher: Check off the first box under Step 1 on the checklist. (*Point to second check box under Step 1.*) What kind of problem is this? How do you know?

Students: An MC problem, because it is comparing the number of kites Michael sold on Saturday to the number of kites he sold on Friday, and the compare words *half as many ... as* in the problem indicate a partial relation between the compared and the referent.

Teacher: Check off the second box under Step 1 on the checklist. Now you are ready for Step 2: Organize the information using the MC diagram. (*Point to first box under Step 2 of checklist.*) To organize the information in an MC problem, what must you do first?

Students: Underline the comparison sentence or question, circle the two things compared, and write them in the diagram.

Teacher: What is the comparison sentence or question in this story? How do you know?

Students: The second sentence, "He sold half as many kites on Saturday as on Friday," is the comparison sentence, because the compare words *half as many ... as* tell about a comparison.

Teacher: Underline this comparison sentence. (*Pause for students to underline.*) What must you do next?

Students: Circle the two things compared and write them in the diagram.

Teacher: What are the two things compared in this problem?

Students: Kites sold on Saturday and kites sold on Friday.

Teacher: Let's circle "Saturday" and "Friday" in the comparison sentence, because they indicate a comparison of kites sold on the two days. (*Pause for students to circle.*) From this comparison sentence, is Saturday or Friday the compared? How do you know?

Students: "Saturday" is the compared, because the number of kites sold on Saturday is compared to the number of kites sold on Friday.

Teacher: Great. So Saturday is the compared and Friday is the referent. Write "Saturday" for the compared and "Friday" for the referent in the diagram. (*Pause for students to complete.*) From this comparison sentence, what is the relation between the number of kites sold on Saturday and on Friday? Is it a multiple or partial relation?

Students: Half, which is a partial relation.

Teacher: Great! Yes, it is a partial relation, because half is a fraction or part of a whole. Circle "half" and write "1/2" for the relation in the diagram. (*Pause for students to circle.*) You underlined the comparison sentence, circled the two things compared, and wrote them in the MC diagram. So cross out the comparison sentence and check off the first box under Step 2 on the checklist. (*Point to second box under Step 2 on checklist.*) What do you do next?

Students: Read the problem to find the numbers for the compared and referent, and write them in the diagram.

Teacher: The first sentence says, "Michael sold 18 kites on Saturday." Does this sentence tell about the compared amount or the referent amount? How do you know?

Students: This sentence tells about the compared amount, because we labeled Saturday (or kites sold on Saturday) as the compared in the diagram.

Teacher: Great! Underline <u>Saturday</u> in this sentence. Circle "18 kites" and write it for Saturday (the compared) in the diagram.

 The last sentence is the question. (*Point to third box under Step 2 on checklist.*) It asks, "How many did he sell on Friday?" Do you know the number of kites sold on Friday?

Students: No.

Teacher: What must you solve for in this problem? (Is it the compared or the referent amount?) How do you know?

Students: The referent amount, because we labeled Friday as the referent in the diagram.

Teacher: Excellent. Underline <u>Friday</u> in the question sentence. Then write a "?" and label (kites) for "Friday" (i.e., the referent) in the diagram. (*Pause for students to complete.*)

 You underlined the compared and the referent, circled numbers and labels (kites), and wrote them in the diagram. You also wrote "? kites" for the referent amount, because you don't know this amount that you need to solve for in this problem. Check off the second and third boxes under Step 2 on the checklist. Now look at the diagram and read what it says.

Students: On Saturday, Michael sold 18 kites. This is only 1/2 as many kites as the amount he sold on Friday. We need to find out the number of kites sold on Friday.

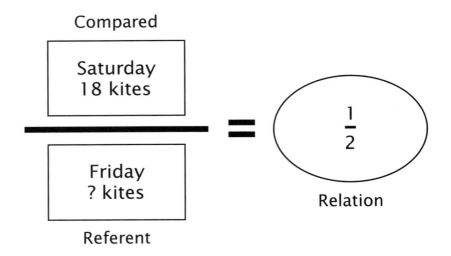

Teacher: Now for Step 3: Plan to solve the problem. (*Point to first box under Step 3 on checklist.*) To plan to solve the problem, what should you do?

Students: Translate the information in the diagram into a math sentence.

Teacher: What is the math sentence when you translate the information in the diagram?

Students: $\frac{18}{?} = \frac{1}{2}$

Teacher: Good. Check off the box under Step 3 on the checklist. Now you are ready for Step 4: Solve the problem. (*Point to first box under Step 4 on the checklist.*) What do you do to solve for the unknown in the equation?

Students: Cross-multiply.

Teacher: What is the answer when you cross-multiply?

Students: $? \times 1 = 18 \times 2$
$? = 18 \times 2$
$? = 36$

Teacher: Check off the first box under Step 4 on the checklist. (*Point to second box under Step 4 on checklist.*) What do you do next?

Students: Write the complete answer on the answer line.

Teacher: The complete answer is the number and the label. What is the complete answer to this MC problem?

Students: 36 kites.

Teacher: Good. Write 36 for the "?" in the diagram and write "36 kites" on the answer line. (*Pause for students to write the answer.*) Check off the second box under Step 4 on the checklist. (*Point to third box under Step 4 on checklist.*) What do you do next?

Students: Check the answer.

Teacher: Does "36 kites" seem right?

Students: Yes, because Michael sold 1/2 as many kites on Saturday (18) as on Friday (36). 18 is 1/2 of 36.

Teacher: Great! You can also check your calculation by redoing the cross-multiplication with all the numbers. That is, $18 \times 2 = 36 \times 1$, or $36 = 36$. So your calculation is correct. (*Check off third box under Step 4.*)

Let's review this MC problem. This problem compared the number of kites Michael sold on Saturday to that on Friday. What's this problem called? Why?

Students: MC, because the comparison sentence tells about a partial relation (i.e., 1/2) between the compared and the referent.

Teacher: Great job working hard. Tomorrow we will practice more MC problems.

Lesson 3: Problem Solution

Materials Needed

Answer Sheet for Paired-Learning Tasks
　Lesson 3: MC Answer Sheet 1

Checklists
　Word Problem–Solving Steps (FOPS) poster
　Multiplicative Compare (MC) Problem–Solving Checklist (laminated copies for students)

Diagram
　Multiplicative Compare Problem diagram poster

Overhead Modeling
　Lesson 3: MC Problems 4 and 5

Student Pages
　Lesson 3: MC Problems 4 and 5
　Lesson 3: MC Worksheet 1

MC Problem 4

Teacher: [*Display Word Problem–Solving Steps (FOPS) poster. Ask students to read each step on the poster that they will use to solve multiplication and division word problems. Display Overhead Modeling page for MC Problem 4. Distribute Student Pages for Problems 4 and 5. Have students look at MC Problem 4.*]

Touch Problem 4. (*Point to MC Problem–Solving Checklist.*) Remember, we will use this checklist that has the same four steps (FOPS) to help us solve MC problems. What's the first step?

Students: Find the problem type.

Teacher: Right. (*Point to the first check box on the MC Problem Checklist.*) To find the problem type, what must you do?

Students: Read the problem and retell it in your own words.

Teacher: Good. I will read the problem aloud.
"Lisa made 12 cupcakes for the Annual Teacher Appreciation week at her school. She made 1/3 as many cupcakes as her friend Mary. How many cupcakes did Mary make?"
I read the problem. What must you do next?

Students: Retell the problem using own words.

Teacher: Yes. Now retell the problem in your own words to help you understand it. (*Call on students to retell the problem. When they retell the problem, remind students to tell what they know in the problem and what they are asked to find out.*)

Students: We know Lisa made 12 cupcakes. We also know that Lisa made 1/3 as many cupcakes as Mary. We don't know how many cupcakes Mary made.

Teacher: Good. Check off the first box under Step 1 on the checklist. (*Point to second check box under Step 1.*) What kind of problem is this? How do you know?

Students: An MC problem, because it is comparing Lisa to Mary on the number of cupcakes made, and the compare words *1/3 as many ... as* in the problem indicate a partial relation between the compared and the referent.

Teacher: Check off the second box under Step 1 on the checklist. Now you are ready for Step 2: Organize the information using the MC diagram. (*Point to first box under Step 2 of checklist.*) To organize the information in an MC problem, what must you do first?

Students: Underline the comparison sentence or question, circle the two things compared, and write them in the diagram.

Teacher: What is the comparison sentence or question in this problem? How do you know?

Students: The second sentence, "She made 1/3 as many cupcakes as her friend Mary," is the comparison sentence, because the compare words *1/3 as many ... as* tell about a comparison.

Teacher: Underline this comparison sentence. (*Pause for students to underline.*) What must you do next?

Students: Circle the two things compared and write them in the diagram.

Teacher: Who is compared to whom in this sentence? What are they compared on?

Students: Lisa is compared to Mary on the number of cupcakes made.

Teacher: Circle "Lisa" and "Mary" in the comparison sentence. (*Pause for students to circle.*) From this comparison sentence, is Lisa or Mary the compared? How do you know?

Students: "Lisa" is the compared, because she is compared to Mary on the number of cupcakes made.

Teacher: Great. So Lisa is the compared, and Mary is the referent. Write "Lisa" in the diagram for the compared and "Mary" for the referent. From this comparison sentence, what is the relation between Lisa and Mary with regard to the number of cupcakes made? Is it a multiple or partial relation?

Students: 1/3, which is a partial relation.

Teacher: Great! Yes, it is a partial relation, because 1/3 is a fraction or part of a whole. Circle "1/3" and write it for the relation in the diagram. (*Pause for students to circle.*) You underlined the comparison sentence, circled the two things compared, and wrote them in the MC diagram. So cross out the comparison sentence and check off the first box under Step 2 on the checklist. *(Point to second box under Step 2 on checklist.)* What do you do next?

Students: Read the problem to find the numbers given for the compared and referent, and write them in the diagram.

Teacher: The first sentence says, "Lisa made 12 cupcakes." Does this sentence tell about the compared amount or the referent amount? How do you know?

Students: This sentence tells about the compared amount, because we labeled Lisa as the compared in the diagram.

Teacher: Great! Underline Lisa in this sentence. Circle "12 cupcakes" and write it for "Lisa" (the compared) in the diagram.

The last sentence is the question. *(Point to third box under Step 2 on checklist.)* It asks, "How many cupcakes did Mary make?" Do you know the number of cupcakes Mary made?

Students: No.

Teacher: What must you solve for in this problem? (Is it the compared or the referent amount?) How do you know?

Students: The referent amount, because we labeled Mary as the referent in the diagram.

Teacher: Excellent. Underline Mary in the question sentence and write a "?" and label (cupcakes) for "Mary" (i.e., the referent) in the diagram, because you don't know this amount. *(Pause for students to complete.)* You underlined the compared and referent, circled numbers and labels (cupcakes), and wrote them in the diagram. You also wrote "? cupcakes" for the referent amount that you need to solve in this problem. Check off the second and third boxes under Step 2 on the checklist. Now look at the diagram and read what it says.

Students: Lisa made 12 cupcakes. 12 cupcakes is only 1/3 as many as the amount Mary made. We need to find out the number of cupcakes Mary made.

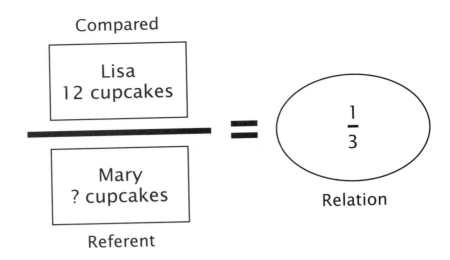

Teacher: Now for Step 3: Plan to solve the problem. (*Point to first box under Step 3 on checklist.*) To plan to solve the problem, what should you do?

Students: Translate the information in the diagram into a math sentence.

Teacher: What is the math equation when you translate the information in the diagram?

Students: $\frac{12}{?} = \frac{1}{3}$

Teacher: Good. Check off the box under Step 3 on the checklist. Now you are ready for Step 4: Solve the problem. (*Point to first box under Step 4 on checklist.*) What do you do to solve for the unknown in the equation?

Students: Cross-multiply.

Teacher: What is the answer when you cross-multiply?

Students: $? \times 1 = 12 \times 3$
$? = 12 \times 3$
$? = 36$

Teacher: Check off the first box under Step 4 on the checklist. (*Point to second box under Step 4 on checklist.*) What do you do next?

Students: Write the complete answer on the answer line.

Teacher: The complete answer is the number and the label. What is the complete answer to this MC problem?

Students: 36 cupcakes.

Teacher: Good. Write 36 for the "?" in the diagram and write "36 cupcakes" on the answer line. (*Pause for students to write the answer.*) Check off the second box under Step 4 on the checklist. (*Point to third box under Step 4 on checklist.*) What do you do next?

Students: Check the answer.

Teacher: Does "36 cupcakes" seem right?

Students: Yes, because Lisa made 1/3 as many cupcakes as Mary. 12 is 1/3 of 36.

Teacher: Great! You can also check your calculation by redoing the cross-multiplication with all the numbers. That is, 36 × 1 = 12 × 3, and 36 = 36. So your calculation is correct. (*Check off third box under Step 4.*)

Let's review this MC problem. This problem compared Lisa to Mary on the number of cupcakes made. What's this problem called? Why?

Students: MC, because the comparison sentence tells about a partial relation (i.e., 1/3) between the compared and the referent.

MC Problem 5

Teacher: (*Display Overhead Modeling page of MC Problem 5. Have students look at MC Problem 5. Point to first check box on MC Problem–Solving Checklist.*) To find the problem type, what must you do?

Students: Read the problem and retell it in own words.

Teacher: Good. I will read the problem aloud.

"Sheila has 5 green beads. Sheila has 30 red beads. How many times as many red beads as green beads does Sheila have?"

I read the problem. Now you retell the problem using your own words.

Students: We know Sheila has 5 green beads. We also know that Sheila has 30 red beads. We do not know the relation (i.e., "n times") between the compared (red beads) and referent (green beads).

Teacher: Great. Check off the first box under Step 1 on the checklist. (*Point to second check box under Step 1.*) What kind of problem is this? How do you know?

Students: An MC problem, because it is comparing Sheila's red beads to her green beads, and the compare words *times as many ... as* in the problem indicate a multiple relation between the compared and the referent.

Teacher: Good. Check off the second box under Step 1 on the checklist. What do you do next?

Students: Organize the information in the problem using the MC diagram.

Teacher: To organize the information in an MC problem, what must you do first? (*Point to first box under Step 2.*)

Students: Underline the comparison sentence or question, circle the two things compared, and write them in the diagram.

Teacher: Good. What is the comparison sentence or question in this problem?

Students: The third sentence, "How many times as many red beads as green beads does Sheila have?" is the comparison question, because the compare words *times as many ... as* tell about a comparison.

Teacher: Underline this comparison question. (*Pause for students to underline.*) What must you do next?

Students: Circle the two things compared and write them in the diagram.

Teacher: What are the two things compared?

Students: Red beads are compared to green beads.

Teacher: Circle "red beads" and "green beads" in the comparison question. (*Pause for students to circle.*) From this comparison question, is the compared item the red beads or the green beads? How do you know?

Students: "Red beads" is the compared, because the number of red beads is compared to the number of green beads.

Teacher: Great. So red beads are the compared, and green beads are the referent. Write "Red beads" in the diagram for the compared and "Green beads" for the referent. From this comparison question, do we know the relation between red and green beads?

Students: No.

Teacher: That is right! We don't know this amount and need to solve for it. Write a "?" for the relation in the diagram. (*Pause for students to complete.*) Check off the first box under Step 2 on the checklist. (*Point to second box under Step 2.*) What do you do next?

Students: Read the problem to find the numbers given for the compared and the referent and write them in the diagram.

Teacher: The first sentence says, "Sheila has 5 green beads." Does this sentence tell about the compared amount or the referent amount? How do you know?

Students: It tells about the referent amount, because we labeled green beads as the referent in the diagram.

Teacher: Great! Underline green beads in this sentence. Circle "5" and write it for "Green beads" (the referent) in the diagram. The second sentence says, "Sheila has 30 red beads." Does this sentence tell about the compared amount or the referent amount? How do you know?

Students: It tells about the compared (red beads) amount, because we labeled red beads as the compared in the diagram.

Teacher: Great! Underline <u>red beads</u> in this sentence, circle "30," and write it for "Red beads" (the compared) in the diagram. (*Pause for students to complete.*)
We underlined the compared and referent, circled numbers, and wrote them in the diagram. We wrote a "?" for the relation that we need to solve for. Check off the second and third boxes under Step 2 on the checklist. Now look at the diagram and read what it says.

Students: Sheila has 30 red beads. Sheila has 5 green beads. We need to find the relation between red beads and green beads (i.e., how many times as many red beads as green beads Sheila has).

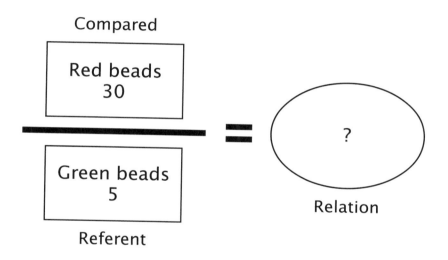

Teacher: Super! Now for Step 3: Plan to solve the problem. (*Point to first box under Step 3 on checklist.*) To plan to solve the problem, what should you do?

Students: Translate the information in the diagram into a math sentence.

Teacher: What is the math equation when you translate the information in the diagram?

Students: $\frac{30}{5} = ?$

Teacher: Good. Check off the first box under Step 3 on the checklist. Now you are ready for Step 4: Solve the problem. (*Point to first box under Step 4 on checklist.*) What is the answer when you solve for the unknown in the equation?

Students: ? = 30/5
? = 6

Teacher: Check off the first box under Step 4 on the checklist. (*Point to second box under Step 4 on checklist.*) What do you do next?

Students: Write the complete answer on the answer line.

Teacher: The complete answer is the number and the label. What is the complete answer to this MC problem?

Students: Sheila has 6 times as many red beads as green beads.

Teacher: Good. Write 6 for the "?" in the diagram and write "6 times as many green beads" on the answer line. (*Pause for students to write the answer.*) Check off the second box under Step 4 on the checklist. (*Point to third box under Step 4 on checklist.*) What do you do next?

Students: Check the answer.

Teacher: Does "6 times as many green beads" seem right? How do you know?

Students: Yes, because Sheila has 30 red beads and 5 green beads; 30 over 5 is 6, or 30 is 6 times of 5.

Teacher: Great! Check off the third box under Step 4. Let's review this MC problem. This problem compared Sheila's red beads to green beads. What's this problem called? Why?

Students: MC, because the comparison question asks for the relation between the red beads and the green beads.
(*Pass out MC Worksheet 1.*)

Teacher: Now I want you to do Problem 1 on this worksheet with your partner.
(*Ask students to* think, plan, *and* share with partners; see Guide to Paired Learning in the Introduction to solve Problem 1 on the worksheet.*)

MC Worksheet 1, Problem 1: "Bill watched 8 Phillies games. He watched 1/4 as many games as Courtney. How many Phillies games did Courtney watch?" (*Use the four steps to solve the MC problem.*)

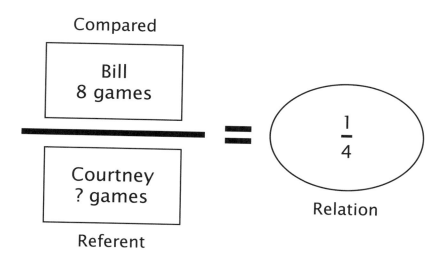

Answer: 32 games

Teacher: Now I want you to do the next two problems on your own. Remember to use the four steps to solve these problems.
   (*Monitor students as they work. Have students check their answers using the MC Answer Sheet 1. Make sure the diagram is labeled correctly, the math equation is written and worked out correctly, the written explanation is complete, and the complete answer is written on the answer line; see below.*)

MC Worksheet 1, Problem 2: "Elaine collected 24 trading cards. Mia collected 2/3 as many cards as Elaine. How many cards did Mia collect?"

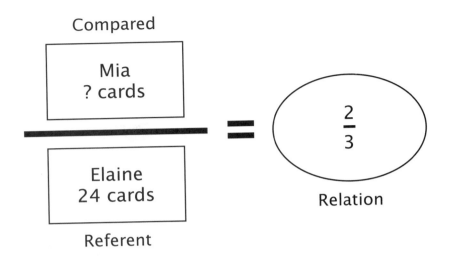

Answer: 16 cards

MC Worksheet 1, Problem 3: "McChicken sold 96 sandwiches every day. It sold 2/3 as many sandwiches as McFast. How many sandwiches did McFast sell every day?"

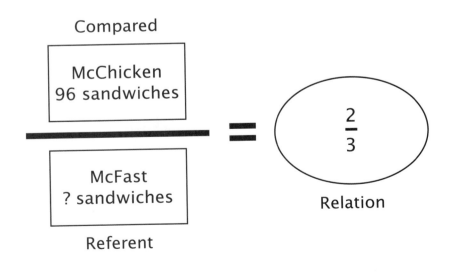

Answer: 144 sandwiches

Teacher: (*Monitor students as they work. After about 10 minutes, go over the answers. Make sure the diagrams are labeled correctly, the math equations are worked out correctly, and the complete answers are written on the answer line.*)

Great job working hard. Tomorrow we will practice more MC problems.

Lesson 4: Problem Solution

Materials Needed

Answer Sheet for Paired-Learning Tasks	Lesson 4: MC Answer Sheet 2
Checklists	Word Problem–Solving Steps (FOPS) poster
	Multiplicative Compare (MC) Problem–Solving Checklist (laminated copies for students)
Diagram	Multiplicative Compare Problem diagram poster
Overhead Modeling	Lesson 4: MC Problem 1
Reference Guide	Lesson 4: MC Reference Guide 2
Student Pages	Lesson 4: MC Worksheet 2

Teacher: (*Pass out MC Worksheet 2. Display Overhead Modeling page of MC Problem 1.*)

Follow along as I read this problem. (*Use guided practice to have students complete Problem 1. Read Problem 1 aloud.*)

"Use the data in the table below to solve MC Worksheet 2, Problem 1."

Name	Weight (pounds)
David	180
Joshua	240
Karen	150
Vivek	140
Freya	60
Carol	?

MC Worksheet 2, Problem 1: "Carol weighs 1/3 as many pounds as David. What is Carol's weight?"

(*See MC Reference Guide 2 for modeling the written explanation. Note: Assist students, if necessary, to identify needed information from the table provided. Because the problem is comparing Carol's weight to that of David, only information related to David and Carol is relevant.*)

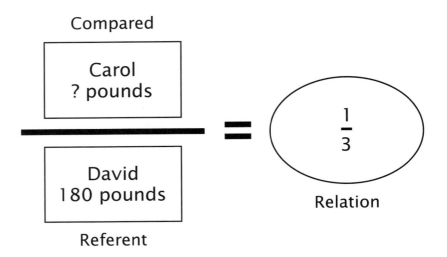

Answer: 60 pounds

Teacher: Now I want you to do the next problem with your partner. (*Ask students to* think, plan, *and* share *with partners; see Guide to Paired Learning in the Introduction to solve Problem 2.*)

MC Worksheet 2, Problem 2: "Patricia spent 4 dollars on a fiction book; she spent 16 dollars on a book bag. How many times as much money did she spend on the book bag as on the fiction book?" (*Use the four steps to solve an MC problem.*)

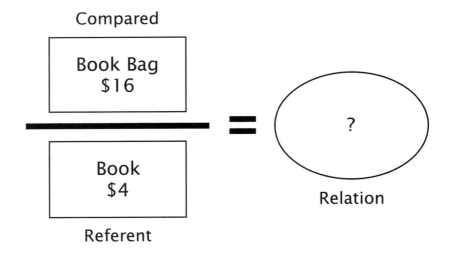

Answer: 4 times

Teacher: Now I want you to do the next four problems on your own. Remember to use the four steps to solve the problems on this worksheet.

(*Monitor students as they work. Have students check their answers using MC Answer Sheet 2. Make sure the diagram is labeled correctly, the math equation is written and worked out correctly, the written explanation is complete, and the complete answer is written on the answer line; see below.*)

"Use the table below to solve MC Worksheet 2, Problem 3."

Day of the Week	Subject	Number of Pages Read
Monday	Social studies	10
	Reading	15
Tuesday	Science	17
	Reading	29
Wednesday	Social studies	23
	Reading	18
Thursday	Social studies	5
	Science	?

MC Worksheet 2, Problem 3: "Joanna writes in a journal where she keeps track of how much she reads each day for different subjects in school. If she read 1/4 as many pages for social studies as science on Thursday, how many pages did Joanna read for science class that day?"

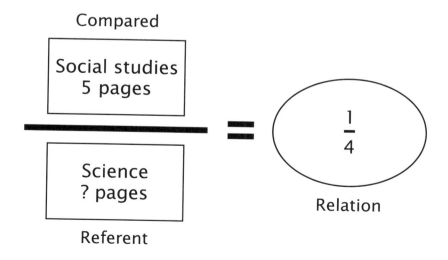

Answer: 20 pages

MC Worksheet 2, Problem 4: "Linda read 12 pages of a children's anthology on Saturday. She read 3 times as many pages on Sunday as on Saturday. How many pages did she read on Sunday?"

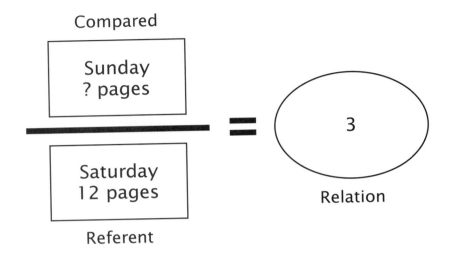

Answer: 36 pages

MC Worksheet 2, Problem 5: "Julie spent $12 at a craft store. She spent 3 times as much as Mike at the store. How much did Mike spend at the craft store?"

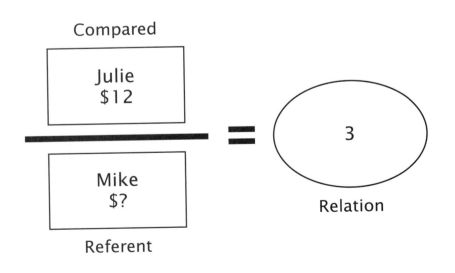

Answer: $4

MC Worksheet 2, Problem 6: "Megan has 135 building blocks in her toy chest. Her brother Ken has 2/3 as many building blocks as Megan. How many building blocks does Ken have?"

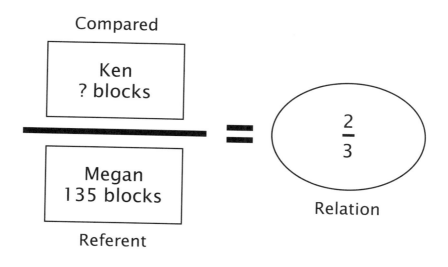

Answer: 90 blocks

Teacher: (*Monitor students as they work. After about 15 to 20 minutes, go over the answers. Make sure the diagrams are labeled correctly, the math equations are written and worked out correctly, and the complete answers are written on the answer line.*)

Great job working hard. Tomorrow you will solve more MC problems using your own diagrams.

Lesson 5: Problem Solution

Materials Needed

Checklists   Word Problem–Solving Steps (FOPS) poster

                      Multiplicative Compare (MC) Problem-Solving Checklist (laminated copies for students)

Overhead Modeling   Lesson 5: MC Problem 1

Student Pages   Lesson 5: MC Worksheet 3

Teacher: (*Pass out MC Worksheet 3. Display Overhead Modeling page of MC Worksheet 3, Problem 1.*)

You learned to solve MC problems using diagrams. Now we will solve the problems on this worksheet using your own diagrams. This worksheet does not have diagrams. Remember to use the four steps (FOPS) to solve problems on the worksheet. (*Note: Discuss how students can generate a diagram that is more efficient than the one they used, and have them practice solving the problems using the diagram they generate. Also, encourage them to use the MC Problem–Solving Checklist only as needed. Use guided practice to complete Problems 1 and 2 using own diagrams; see below.*)

MC Worksheet 3, Problem 1: "Betty and Sara both teach mathematics in Liberty High School. Betty has 27 students in her class. She has 3 times as many students as Sara. How many students does Sara have?"

$$\frac{\text{Betty: 27 students}}{\text{Sara: ? students}} = 3$$

Answer: 9 students

MC Worksheet 3, Problem 2: "Jennie got 20 Easter bunny candies. Tina got 4/5 as many Easter bunny candies as Jennie. How many Easter bunny candies did Tina get?"

$$\frac{\text{Tina: ? candies}}{\text{Jennie: 20 candies}} = \frac{4}{5}$$

Answer: 16 candies

Teacher: Now I want you to do the next three problems on your own. (*Have students write the explanation for at least one of the three problems.*) Remember to use the four steps to solve Problems 3 through 5 on the worksheet.

MC Worksheet 3, Problem 3: "Gillian picked 24 apples during her trip to the Green Garden. Sue picked 3 times as many as Gillian. How many apples did Sue pick?"

$$\frac{\text{Sue: ? apples}}{\text{Gillian: 24 apples}} = 3$$

Answer: 72 apples

MC Worksheet 3, Problem 4: "Carlos ate 6 steamed dumplings at the East Chinese restaurant last weekend. He ate 2/3 as many dumplings as his dad. How many dumplings did his dad eat?"

$$\frac{\text{Carlos: 6 dumplings}}{\text{Dad: ? dumplings}} = \frac{2}{3}$$

Answer: 9 dumplings

MC Worksheet 3, Problem 5: "Jennie has 18 red markers and 4 purple markers. How many times as many red markers as purple markers does Jennie have?"

$$\frac{\text{Red: 18 markers}}{\text{Purple: 4 markers}} = ?$$

Answer: 4.5 times

Teacher: You learned to solve MC word problems using your own diagrams. Next you will learn to map information in vary word problems onto diagrams and solve them.

Unit 2

Vary Problems

Lesson 6: Problem Schema

Materials Needed

Checklist	Vary Story Checklist (laminated copies for students)
Diagram	Vary Problem diagram poster
Overhead Modeling	Lesson 6: Vary Stories 1, 2, and 3
Student Pages	Lesson 6: Review Worksheet 1
	Lesson 6: Vary Stories 1, 2, and 3
	Lesson 6: Vary Schema Worksheet 1

Teacher: (*Pass out Review Worksheet 1. Review solving Multiplicative Compare problems using Problems 1 and 2 on Review Worksheet 1.*)

Now we will learn to identify and organize another type of multiplication and division problem called "vary" so that we can later solve it. A vary problem tells about an association (ratio or rate) between two things. Often there is an *if–then* kind of statement that tells about two pairs of association between two things.

For example, "*If* one pencil costs 8 cents, *then* a pack of 12 pencils will cost 96 cents." In this story, the two things that form a specific ratio are *pencils* and *cents*. That is, one pencil costs 8 cents. And there is an *if–then* statement in the story: "If one pencil costs 8 cents, then a pack of 12 pencils will cost 96 cents." It is important to understand that the ratio (i.e., 1 pencil/8 cents) formed in the *if* statement and the ratio (i.e., 12 pencils/96 cents) formed in the *then* statement are equivalent or proportional. That is, 1 pencil/8 cents = 12 pencils/96 cents (if we reduce 12/96 to the lowest term, we will get 1/8). (*Display Vary Story Checklist.*)

Here are two steps we will use to organize information in a vary story. (*Point to and read each step on the Vary Story Checklist.*)

Vary Story 1

Teacher: Let's use these two steps to do an example. Look at this story. (*Display Overhead Modeling page of Vary Story 1. Pass out student copies* of Vary Stories 1, 2, and 3 and Vary Story Checklist. Point to first check box on Vary Story Checklist)*. Now we are ready for Step 1: Find the problem type. To find the problem type, I will read the story and retell it in my own words. (*Read story aloud.*)

"If there are 8 seats in each rowboat, then 5 rowboats will have 40 seats."

Now I'll retell the problem in my own words to help me understand it. When I retell, I will ask myself, What do I know in this story. (*Retell the story.*)

This story tells about an association or a ratio between rowboats and the number of seats in rowboats. If each rowboat has 8 seats in it, then 5 rowboats will have a total of 40 seats.

I read the story and told it in my own words. Let's check off the first box under Step 1 on the checklist. (*Point to second check box under Step 1.*) Now I will ask myself if the story is a vary problem type. How do I know it is a vary problem? I will ask myself whether the story tells about a ratio or rate between two things and whether there is an *if–then* statement.

In this story, the *if* statement tells about a ratio between one rowboat and the number of seats (8) in it. This, in fact, is a unit ratio. The *then* statement in the story also tells about the association between rowboats (5) and seats (40). So this is a vary story, because it tells about the association (or ratio) between the number of rowboats and seats in rowboats. (*Check off second box under Step 1.*).

Now I am ready for Step 2: Organize the information in the story using the vary diagram. (*Display Vary Problem diagram poster*).

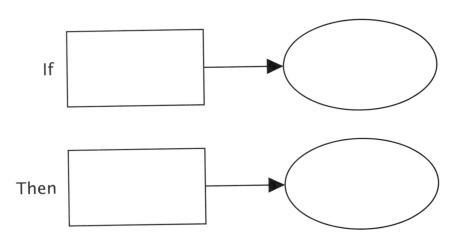

Vary Problem

(*Point to first box under Step 2.*) To organize the information in a vary problem, we first underline the two things that form a specific ratio and write their names in the diagram. What are the two things that form a specific ratio in this story? How do you know?

Students: Rowboats and seats in rowboats, because both the *if* and the *then* statements talk about the association between rowboats and seats.

Teacher: That is right. Both statements talk about rowboats and seats. That is, the story tells about two pairs of associations between rowboats and seats. Let's underline <u>rowboats</u> and <u>seats</u> in the *if–then* statements and write "Rowboats" and "Seats in rowboats" as labels for the two columns in the vary diagram. (*Point out that it is okay to write either rowboats or seats in the first column; however, it is important to align the numbers for rowboats and seats in columns. Pause for students to write.*)

Let's check off the first box under Step 2 on the checklist. (*Point to second box under Step 2.*) Next we need to read the story to circle numbers for each pair of association and write numbers and labels in the diagram. The first part of the sentence says, "If there are 8 seats in *each* rowboat." This is the *if* statement. Let's write the numbers and labels for the first pair of association (i.e., the *if* statement) in the diagram. (*Point to row for* if *statement in diagram.*) What does *each* in the sentence mean?

Students: *Each* means "one."

Teacher: Great! That is, 8 seats are in one rowboat, or one rowboat has 8 seats. Let's circle "each" and write "1 rowboat" in the box for the "Rowboats" column. Also, let's circle "8" and write "8 seats" in the oval for the "Seats" column. (*Pause for students to complete.*) The second part of the sentence says, "then 5 rowboats will have 40 seats." This is the *then* statement, which indicates that 5 rowboats have 40 seats, or 40 seats are in 5 rowboats.

Now we are ready to write the numbers for the second pair of association (the *then* statement) in the diagram. (*Point to second row for* then *statement in diagram.*) Let's circle "5" and write "5 rowboats" in the box for the "Rowboats" column. Also, circle "40" and write "40 seats" in the oval for the "Seats" column. (*Explain that because we named the first column "Rowboats," we need to write "5 rowboats" in the box, and because we named the second column "Seats," we write "40 seats" in the oval. Pause for students to complete.*)

We underlined the two things (<u>rowboats</u> and <u>seats</u>) in the story that formed an association and wrote their names in the two columns in the diagram, circled numbers, and wrote numbers and labels in the vary diagram. Let's check off the second box under Step 2 on the checklist. Now let's look at the diagram and read what it says.

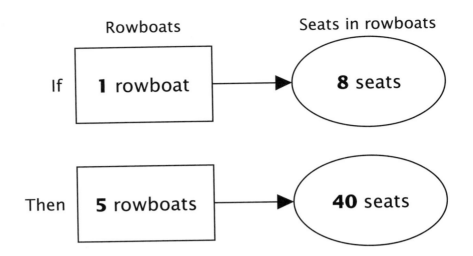

(*Point to relevant parts of diagram as you explain.*) If one rowboat has 8 seats in it, then 5 rowboats have 40 seats in them. Remember, it is very important to write numbers for rowboats in the "Rowboats" column and number of seats in the "Seats" column in the diagram. This diagram also shows that the ratio (1 to 8, or 1/8) presented in the *if* statement and the ratio (5 to 40, or 5/40) presented in the *then* statement are equivalent or proportional. That is, 1/8 = 5/40; if we reduce 5/40 to the lowest term, we will get 1/8. Does this make sense to you?

Students: This seems to make sense, because if 1 rowboat has 8 seats, then 5 would have 5 times 8 or 40 seats.

Teacher: That's right. This is a vary problem, because it tells about an association (i.e., ratio or rate) between two things, and because the *if–then* statements make up two pairs of associations or two ratios, and the two ratios are equivalent.

Vary Story 2

Teacher: (*Display Overhead Modeling page of Vary Story 2. Have students look at Vary Story 2.*)
Touch Story 2. (*Point to checklist.*) What's the first step?

Students: Find the problem type.

Teacher: (*Point to first check box on Vary Story Checklist.*) To find the problem type, I will read the story and retell it in my own words. (*Read story aloud.*)
"If a school bus holds 50 students, then 26 school buses are needed to bus 1,300 students to their school."
I read the story. What must I do next?

Students: Retell the story using own words.

Teacher: Yes, I will retell the story in my own words to help me understand it. When I retell, I will ask myself, What do I know in this story? (*Retell the problem.*)

This story tells about an association or a ratio between school buses and students. That is, if one school bus holds 50 students, then 26 school buses will hold 1,300 students.

I read the story and told it in my own words. Let's check off the first box under Step 1 on the checklist. (*Point to second check box under Step 1.*) What do I do next?

Students: Ask if the story is a vary problem type.

Teacher: Good. How do I know if it is a vary problem? I will ask whether the story tells about a ratio or rate between two things—that is, whether there is an *if–then* statement. In this story, the *if* statement tells about a ratio between one bus and the number of students (50) it can hold, which is a unit ratio. The *then* statement in the story also tells about the association between buses and the number of students. That is, 26 buses will hold 1,300 students. So this story is a vary story that tells about the association (or rate) between buses and the number of students. (*Check off second box under Step 1.*)

Now I am ready for Step 2: Organize the information in the story using the vary diagram. (*Display Vary Problem diagram poster.*)

Vary Problem

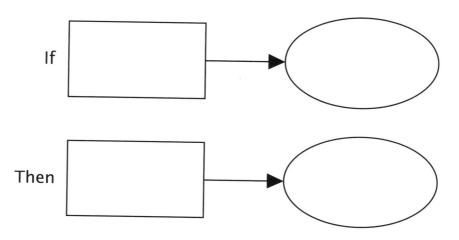

(*Point to first box under Step 2.*) To organize the information in a vary problem, we first underline the two things that form a specific ratio and write their names in the diagram. What are the two things that form a specific ratio in this story? How do you know?

Students: Buses and students, because both the *if* and the *then* statements talk about the association between buses and students.

Teacher: That is right. Both statements talk about buses and students. That is, the story tells about two pairs of associations between buses and students. Let's underline <u>buses</u> and <u>students</u> in the *if–then* statements and write "Buses" and "Students" as labels for the two columns in the vary diagram. (*Pause for students to write.*) Let's check off the first box under Step 2 on the checklist. (*Point to second box under Step 2.*)

Next we need to read the story to circle numbers for each pair of association and write numbers and labels in the diagram. The first part of the sentence says, "If a school bus holds 50 students." This is the *if* statement. Let's write the numbers and labels for the first pair of association (i.e., the *if* statement) in the diagram. (*Point to the row for the* if *statement in the diagram.*) What does "A school bus" mean?

Students: "A school bus" means "one school bus."

Teacher: Great! One school bus holds 50 students. Let's circle "a" and write "1 bus" in the box for the "Buses" column. Also, let's circle "50" and write "50 students" in the oval for the "Students" column. (*Pause for students to complete.*)

The second part of the sentence says, "then 26 school buses are needed to bus 1,300 students to their school." This is the *then* statement, which indicates that 26 school buses hold 1,300 students. Now we are ready to write the numbers for the second pair of association (the *then* statement) in the diagram. (*Point to the second row for the* then *statement in the diagram.*) Let's circle "26" and write "26 buses" in the box for the "buses" column. Also, circle "1,300" and write "1,300 students" in the oval for the "Students" column. (*Pause for students to complete.*)

We underlined the two things (<u>buses</u> and <u>students</u>) in the story that formed an association and wrote their names in the two columns in the diagram, circled numbers, and wrote numbers and labels in the vary diagram. Let's check off the second box under Step 2 on the checklist. Now let's look at the diagram and read what it says.

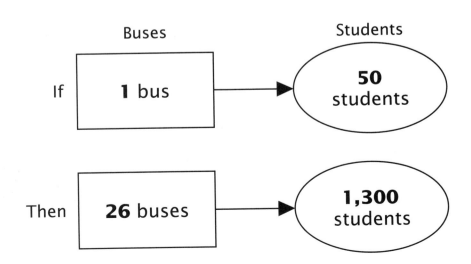

(*Point to relevant parts of the diagram as you explain.*) If one bus holds 50 students, then 26 buses hold 1,300 students. Remember, it is very important to write numbers for buses in the "Buses" column and the number of students in the "Students" column in the diagram. This diagram also shows that the ratio (1 to 50, or 1/50) presented in the *if* statement and the ratio (26 to 1,300, or 26/1,300) presented in the *then* statement are equivalent or proportional. That is, 1/50 = 26/1,300; if we reduce 26/1,300 to the lowest term, we will get 1/50. Does this make sense to you?

Students: This seems to make sense, because if 1 bus holds 50 students, then 26 buses would hold 26 times 50 or 1,300 students.

Teacher: That's right. This is a vary problem, because it tells about an association (i.e., ratio or rate) between two things; and the *if–then* statements make up two pairs of associations or two ratios, and the two ratios are equivalent.

Vary Story 3

Teacher: (*Use the script as a guideline for mapping information in Vary Story 3, and facilitate understanding and reasoning by having frequent student–teacher exchanges. Display Overhead Modeling page of Vary Story 3. Have students look at Vary Story 3.*)

Touch Story 3. (*Point to first check box on Vary Story Checklist.*) What's the first step? What do we need to do?

Students: Find the problem type by reading the story and retelling it.

Teacher: Great! Read the story aloud (or have a student read it).
"Chris and Dianne are responsible for making drinks for the party. They used 5 lemons to make 2 quarts of lemonade. To make 8 quarts of lemonade, they need 20 lemons."
You read the problem. What must you do next?

Students: Retell the story using own words.

Teacher: Yes. Retell the story in your own words to help you understand it. (*Call on students to retell the story. When they retell the story, remind them to tell what they know in the story.*)

Students: This story tells about an association between lemons and lemonade. One sentence in the story tells us that 5 lemons are used to make 2 quarts of lemonade. Another sentence says they need 20 lemons to make 8 quarts of lemonade.

Teacher: Check off the first box under Step 1 on the checklist. (*Point to second check box under Step 1.*) What kind of problem is this? What must you look for to find out whether it is a vary problem?

Students: The story should tell about a ratio or rate between two things and involve an *if–then* statement.

Teacher: Great! Does the story tell about a ratio or rate between two things?

Students: Yes, the story tells about a ratio between lemons and lemonade.

Teacher: Is there an *if–then* statement in the story?

Students: No.

Teacher: That is right! In this story, although we do not see the words *if–then,* there is an *if–then* relation (not directly stated) that tells about two pairs of association between lemons and lemonade. That is, "5 lemons are used to make 2 quarts of lemonade" is the first pair of lemon–lemonade association, and "20 lemons are needed to make 8 quarts of lemonade" is the second pair of association. We can assume the first pair of association to be the *if* statement and the second pair of association to be the *then* statement. That is, "*if* 5 lemons are used to make 2 quarts of lemonade, *then* 20 lemons are needed to make 8 quarts of lemonade." So can we say that this is a vary story?

Students: Yes.

Teacher: Super! Check off the second box under Step 1. Now for Step 2: Organize the information in the story using the vary diagram. (*Display Vary Problem diagram poster.*)

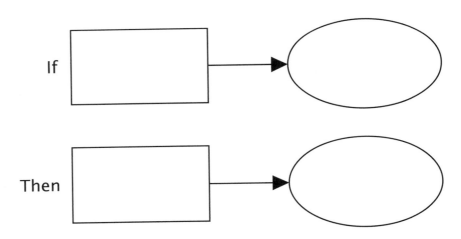

(*Point to first box under Step 2.*) To organize the information in a vary problem, we first underline the two things that form a specific ratio or rate and write their names in the diagram. What does this story talk about?

Students: Lemons and lemonade.

Teacher: Great! Does the story describe a specific ratio between lemons and lemonade? How do you know?

Students: Yes, because the second and third sentences in the story talk about the number of lemons needed to make quarts of lemonade.

Teacher: Good! Underline lemons and lemonade in the story and write "Lemons" for the first column and "Lemonade" for the second column in the diagram. (*Pause for students to complete.*) Check off the first box under Step 2 on the checklist. (*Point to second box under Step 2.*)

Now read each sentence, circle the numbers for each of the two pairs of associations, and write the numbers and labels in the diagram. What does the first sentence say?

Students: "Chris and Dianne are responsible for making drinks for the party."

Teacher: Does this sentence tell about any one of the pairs of association?

Students: No.

Teacher: Let's cross it out, because you don't need this information to solve the problem. The second sentence says, "They used 5 lemons to make 2 quarts of lemonade." Does this sentence tell about one pair of association? How do you know?

Students: Yes, it tells that 5 lemons make 2 quarts of lemonade.

Teacher: So what numbers do you circle and write for the *if* statement in the diagram?

Students: We circle "5" and write "5 lemons" in the box for the "Lemons" column and circle "2" and write "2 quarts of lemonade" in the oval for the "Lemonade" column.

Teacher: Great! (*Pause for students to complete.*) The last sentence says, "To make 8 quarts of lemonade, they need 20 lemons." Does this sentence tell about the second pair of association? How do you know?

Students: Yes, it tells that to make 8 quarts of lemonade, you need 20 lemons.

Teacher: Right! We can also restate it as 20 lemons make 8 quarts of lemonade. Now you need to circle and write the numbers and labels for this second pair of association (the *then* statement) in the diagram. Do you write "8 quarts" in the box or oval for the *then* statement in the diagram? How do you know?

Students: We write 8 quarts in the oval for the "Lemonade" column, because the box is for the number of lemons not lemonade.

Teacher: Super! Write "8 quarts" in the oval for the "Lemonade" column. (*Pause for students to write.*) What number and label do we write in the box for the "Lemon" column?

Students: 20 lemons.

Teacher: Great! Write "20 lemons" in the box. (*Pause for students to write.*) You underlined the two things (lemons and lemonade) in the story that formed an association. You also wrote their names in the two columns in the diagram, circled numbers, and wrote numbers and labels in the vary diagram. Check off the second box under Step 2. Now let's look at the diagram and read what it says.

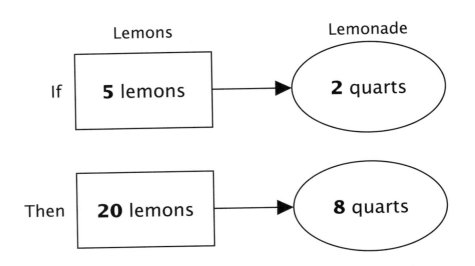

(*Point to relevant parts of the diagram as you explain.*) If 5 lemons are needed for 2 quarts of lemonade, then 20 lemons are needed for 8 quarts of lemonade. Because this is a vary story, the ratio (5 to 2 or 5/2) presented in the *if* statement and the ratio (20 to 8, or 20/8) presented in the *then* statement are equivalent or proportional. That is, 5/2 = 20/8, or 5/20 = 2/8 (if we reduce 5/20 to the lowest term, and 2/8 to the lowest term, we will get 1/4 = 1/4). Does this make sense to you?

Students: This seems to make sense, because if 5 lemons make 2 quarts of lemonade, 20 lemons will make 4 times of 2 quarts, which would be 8 quarts.

Teacher: Great! What is this story called? Why?

Students: Vary, because it tells an association (i.e., ratio) between lemons and lemonade, and because there are two pairs of associations that make up an *if–then* statement.

Teacher: Great! And the *if–then* statements in the diagram make up two equivalent ratios between lemons and lemonade.

(*Pass out Vary Schema Worksheet 1.*)

Now I want you to do the next five stories on your own. Remember to use the two steps to organize information in stories using the vary diagram.

(*Monitor students as they work. Then, check the information in the diagrams. Make sure the diagrams are labeled correctly and completely; see below.*)

Vary Schema Story 1: "If one circus ticket costs $5, then $30 can buy 6 circus tickets."

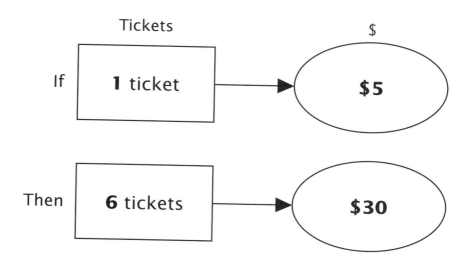

Vary Schema Story 2: "Each skating board has 4 wheels; 12 skating boards will have 48 wheels."

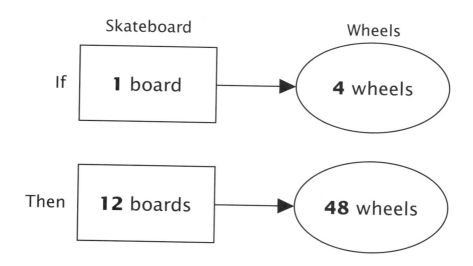

Vary Schema Story 3: "If Mr. Smith's car will run 21 miles on each gallon of gas, then 4 gallons of gas will allow the car to drive 84 miles."

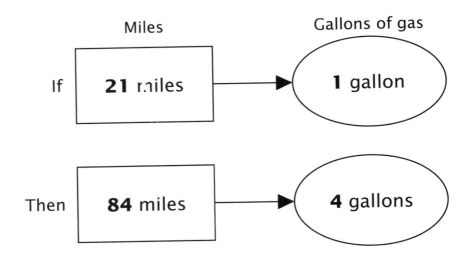

Vary Schema Story 4: "If a pack of socks costs $5, then 7 packs of socks will cost $35."

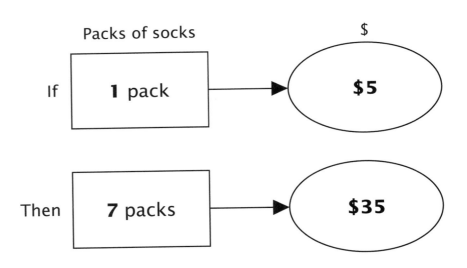

Vary Schema Story 5: "In a classroom, each row can only seat 6 students. There are 24 students seated in 4 rows."

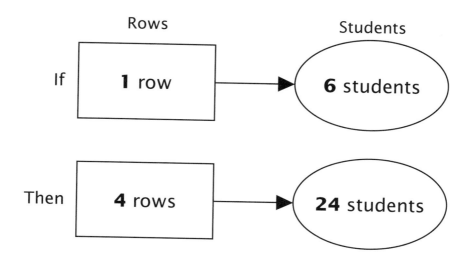

Teacher: You learned to map information in vary story situations onto diagrams. Next you will learn to solve vary problems.

Lesson 7: Problem Solution

Materials Needed

Checklists	Word Problem–Solving Steps (FOPS) poster
	Vary Problem–Solving Checklist (laminated copies for students)
Diagram	Vary Problem diagram
Reference Guide	Lesson 7: Vary Reference Guide 1
Overhead Modeling	Lesson 7: Vary Problems 1, 2, and 3
Student Pages	Lesson 7: Vary Problems 1, 2, and 3

Teacher: Today we are going to use vary diagrams like the ones you learned earlier to solve vary word problems. Let's review the vary problem. A vary problem tells about an association (ratio or rate) between two things. There is an *if–then* kind of statement that makes up two pairs of association between two things, and they are equivalent ratios.

[*Display Word Problem-Solving Steps* (FOPS) *poster.*]

Do you remember the funny word FOPS? What are the four steps in FOPS?

Students: F—Find the problem type; O—Organize the information using a diagram; P—Plan to solve the problem; S—Solve the problem.

Vary Problem 1

Teacher: (*Display Overhead Modeling page of Vary Problem 1, and see Vary Reference Guide 1 to set up the problem. Pass out copies of Vary Problem 1 and Vary Problem–Solving Checklist.*)

(*Point to Vary Problem Checklist.*) We will use this checklist that has the same four steps (FOPS) to help us solve vary problems.

We are ready for Step 1: Find the problem type. (*Point to first check box on Vary Problem–Solving Checklist.*) To find the problem type, I will read the problem and retell it in my own words. Follow along as I read Problem 1. (*Read Vary Problem 1 aloud.*)

"The Joy Company packs 48 cans of tomatoes in each crate. How many crates will the company need to pack 720 cans of tomatoes?"

Now I'll retell the problem in my own words to help me understand it. When I retell, I will ask myself, What do I know in this problem and what am I asked to find out? (*Retell the problem.*)

This story tells about an association or a ratio between the number of cans of tomatoes and the number of crates. I know 48 cans of tomatoes are packed in each crate; I do not know how many crates will be needed to pack 720 cans of tomatoes, which is what I am asked to find out.

I read the story and told it in my own words. Let's check off the first box under Step 1 on the checklist. (*Point to second check box under Step 1.*) Now I will ask myself if the story is a vary problem type. To find out if this is a vary problem, I will ask myself whether the problem tells about a ratio or rate between two things and whether there is an *if–then* statement. Does this problem tell about a rate or ratio between two things?

Students: Yes, the problem tells about a ratio between the number of cans of tomatoes and the number of crates.

Teacher: Is there an *if–then* statement directly stated in this problem?

Students: No.

Teacher: Although we do not see the words *if* and *then,* this problem describes an *if–then* statement that tells about two pairs of association between the number of cans of tomatoes and the number of crates needed to pack the cans. The first sentence tells a specific ratio between the number of cans of tomatoes and the number of crates (i.e., 48 cans of tomatoes to each crate); the second sentence is a question that asks about the ratio between 720 cans of tomatoes and an unknown number of crates. That is, both the sentence and question talk about the association between number of cans of tomatoes and number of crates. We use the ratio provided in the first sentence (i.e., the *if* statement) to solve for the number of crates needed to pack 720 cans of tomatoes. Typically, we assume the statement that describes a specific ratio to be the *if* statement and use that information to find the unknown quantity in the other pair of association. That is, we can restate the problem as follows: *If* "the Joy Company packs 48 cans of tomatoes in each crate," *then* "how many crates will the company need to pack 720 cans of tomatoes?" So can we say that this is a vary story?

Students: Yes.

Teacher: Great. Let's check off the second box under Step 1. Now I am ready for Step 2: Organize the information in the problem using the vary diagram. (*Display Vary Problem diagram.*)

Vary Problem

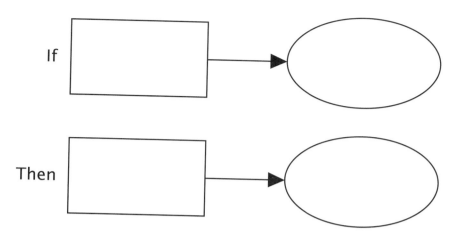

(*Point to first box under Step 2.*) To organize the information in a vary problem, we first underline the two things that form a specific ratio and write their names in the diagram. What does this problem talk about?

Students: Cans of tomatoes and crates.

Teacher: Great! Does the problem describe a specific ratio between cans of tomatoes and crates? How do you know?

Students: Yes, because both the statement and the question in the problem talk about the number of cans of tomatoes and the number of crates needed to pack cans of tomatoes.

Teacher: Good! Underline cans of tomatoes and crate(s) in the problem and write "Cans of tomatoes" for the first column and "Crate(s)" for the second column in the diagram. (*Pause for students to complete.*) Check off the first box under Step 2 on the checklist. (*Point to second box under Step 2 of checklist.*) Now read each sentence to circle numbers for each of the two pairs of associations and write numbers and labels in the diagram. What does the first sentence say?

Students: "The Joy Company packs 48 cans of tomatoes in each crate."

Teacher: What does "each" means?

Students: "Each" means "one."

Teacher: So what numbers do you circle and write for the *if* statement in the diagram?

Students: We circle "48" and write "48 cans" in the box for the "Cans of tomatoes" column and circle "each" and write "1 crate" in the oval for the "Crates" column.

Teacher: Great! (*Pause for students to complete.*) The next sentence is a question. It asks, "How many crates will the company need to pack

720 cans of tomatoes?" Now you need to circle and write numbers and labels for the *then* statement in the diagram. From the question sentence, do you know the number of cans of tomatoes to write in the *then* part of the vary diagram?

Students: Yes, 720 cans of tomatoes.

Teacher: Do you write "720 cans" in the box or oval for the *then* statement in the diagram? How do you know?

Students: Write in the box, because it tells about the number of cans of tomatoes.

Teacher: Great! Circle "720" and write "720 cans" in the box. (*Pause for students to write.*) What number and label do we write in the Oval for the "Crates" column? Do you know the number for the Crates column?

Students: No. This is what we are asked to solve.

Teacher: Correct. We do not know how many crates are needed to pack 720 cans of tomatoes. Let's write "? crates" in the oval for the "crates" column. (*Pause for students to complete.*) We underlined the two things (cans of tomatoes and crates) in the problem that formed a ratio and wrote their names in the two columns in the diagram, circled numbers, and wrote numbers and labels for what we know and wrote a "?" for what must be solved in the vary diagram. Check off the second and third boxes under Step 2. Now let's look at the diagram and read what it says.

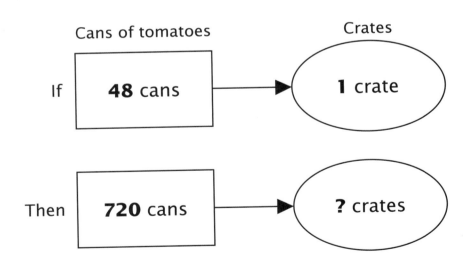

(*Point to relevant parts of the diagram as you explain.*) If 48 cans of tomatoes can be packed in 1 crate, then 720 cans of tomatoes can be packed in "?" crates. We need to find out the number of crates that can pack 720 cans of tomatoes. What must you solve for in this problem?

Students: Number of crates to pack 720 cans of tomatoes in the second pair of association in the diagram.

Teacher: Great! Now for Step 3: Plan to solve the problem. (*Point to first box under Step 3 on checklist.*) To plan to solve the problem, we need to translate the information in the diagram into a math equation. In the vary problem, the ratio presented in the *if* statement (48 cans to 1 crate, or 48/1) and the ratio presented in the *then* statement (720 cans to ? crates, or 720/?) are equivalent or proportional. The question can be set up in either of two ways:

Equation A

$$\frac{48 \text{ cans}}{1 \text{ crate}} = \frac{720 \text{ cans}}{? \text{ crates}}$$

OR

Equation B

$$\frac{48 \text{ cans}}{720 \text{ cans}} = \frac{1 \text{ crate}}{? \text{ crates}}$$

(*Use cross-multiplication to show that Equation A and Equation B are the same.*) Equation A is the same as equation B. However, because the mapping of the numbers in Equation B is the same as in the vary diagram, we will use this equation to plan to solve the problem. That is,

$$\frac{48}{720} = \frac{1}{?}$$

Let's check off the first box under Step 3 on the checklist. Now we are ready for Step 4: Solve the problem. (*Point to first box under Step 4 on checklist.*) We need to solve for the unknown or "?" in the equation. We will solve for the "?" by applying the cross-multiplication method. That is,

$$48 \times ? = 720 \times 1$$

$$? = 720 \div 48$$

$$? = 15$$

Now check off the first box under Step 4 on the checklist. (*Point to second box under Step 4 on checklist.*) Next write 15 for the "?" in the diagram and write the complete answer on the answer line. The complete answer is the number and the label. What is the complete answer to this vary problem?

Students: 15 crates.

Teacher: Good. Write 15 crates on the answer line. (*Pause for students to write the answer.*) Let's check off the second box under Step 4 on the checklist. (*Point to third box under Step 4 on checklist.*) We are now ready to check the answer. Does "15 crates" seem right?

Yes, because if one crate packs 48 cans of tomatoes, then 15 crates will pack 15 times of 48; that is, $15 \times 48 = 720$. So our answer (i.e., 15 crates) seems right. I can also check by redoing the cross-multiplication using all the numbers. That is, $48 \times 15 = 720 \times 1$, or $720 = 720$. So my calculation is correct. (*Check off third box under Step 4.*)

(*Model writing the explanation for how the problem was solved here; see Vary Reference Guide 1.*) Let's review this vary problem. This problem tells about an association (i.e., ratio) between cans of tomatoes and number of crates. The *if–then* statements make up two pairs of associations or two ratios, and the two ratios are equivalent.

Vary Problem 2

Teacher: (*Display Overhead Modeling page of Vary Problem 2. Have students look at Vary Problem 2.*)

Touch Vary Problem 2. (*Point to Vary Problem–Solving Checklist.*) What's the first step?

Students: Find the problem type.

Teacher: Right. (*Point to first check box on checklist.*) To find the problem type, I will read the problem and retell it in my own words. (*Read the problem aloud.*)

"Ms. Baker made 60 almond cookies using 8 eggs. If Ms. Baker has only 2 eggs, how many almond cookies can she make?"

I read the problem. What must I do next?

Students: Retell the problem using own words and discover the problem type.

Teacher: Yes, I will retell the problem in my own words to help me understand it. When I retell, I will ask myself, What do I know in this problem and what am I asked to find out? (*Retell aloud.*)

This story tells about an association or a ratio between the number of eggs and number of almond cookies (that can be made from eggs). I know Ms. Baker made 60 almond cookies using 8 eggs, or 8 eggs can make 60 almond cookies; I do not know how many almond cookies can be made from 2 eggs.

I read the story and told it in my own words. Let's check off the first box under Step 1 on the checklist. (*Point to second check box under Step 1.*) Now I will ask myself if the story is a vary problem type.

To find out if this is a vary problem, I will ask myself whether the problem tells about a ratio or rate between two things and whether there is an *if–then* statement. Does this problem tell about a rate or ratio between two things?

Students: Yes, the problem tells about an association or a ratio between the number of cookies and the number of eggs needed to make cookies.

Teacher: Is there an *if–then* statement in this problem?

Students: Yes, even though it is not directly stated in the problem.

Teacher: Let's see, the first sentence tells about a ratio between the number of cookies that can be made and the number of eggs needed to make them. The second sentence, which is the question, asks how many almond cookies can be made from 2 eggs. Therefore, both the sentence and the question talk about the association between the number of cookies and the number of eggs. Typically, we assume the statement that describes a specific ratio to be the *if* statement and use that information to find the unknown quantity in the other pair of association. We can restate the problem as follows: If Ms. Baker made 60 almond cookies using 8 eggs, then 2 eggs can make how many cookies? Is this a vary problem? How do you know?

Students: Yes. Because the problem tells about a ratio between the number of cookies that can be made and the number of eggs needed to make the cookies, and because there is an *if–then* statement.

Teacher: Great! Let's check off the second box under Step 1. Now I am ready for Step 2: Organize the information in the problem using the vary diagram. (*Display Vary Problem diagram poster.*)

Vary Problem

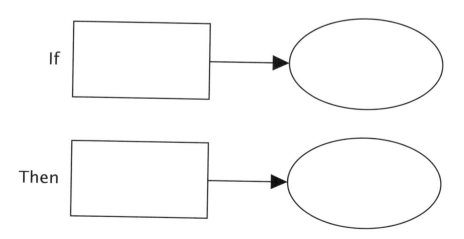

(*Point to first box under Step 2.*) To organize the information in a vary problem, we first underline the two things that form a specific ratio and write their names in the diagram. What does this problem talk about?

Students: Cookies and eggs.

Teacher: Great! Does the problem describe a specific ratio between cookies and eggs? How do you know?

Students: Yes, because both the statement and the question in the problem talk about the number of cookies and the number of eggs needed to make cookies.

Teacher: Good! Underline <u>cookies</u> and <u>eggs</u> in the problem, and write "cookies" for the first column and "eggs" for the second column in the diagram. (*Pause for students to complete.*) Check off the first box under Step 2 on the checklist. (*Point to second box under Step 2 of checklist.*) Now read each sentence to circle the numbers for each of the two pairs of associations and write the numbers and labels in the diagram. What does the first sentence say?

Students: Ms. Baker made 60 almond cookies using 8 eggs.

Teacher: What numbers do you circle and write for the *if* statement in the diagram?

Students: We circle "60" and write "60 cookies" in the box for the "Cookies" column and circle "8" and write "8 eggs" in the oval for the "Eggs" column.

Teacher: Great! (*Pause for students to complete.*) Now you need to circle and write numbers and labels for the *then* statement in the diagram. The question asks, "If Ms. Baker has only 2 eggs, how many almond cookies can she make?" Do you know the number for the "cookies" in the *then* part of the diagram?

Students: No. This is what we are asked to solve.

Teacher: Correct. We do not know how many cookies can be made with 2 eggs. Let's write "? cookies" in the box for the "Cookies" column. (*Pause for students to complete.*) Do you know the number for the "eggs" in the *then* part of the diagram?

Students: Yes, 2 eggs.

Teacher: Great! Circle "2" and write "2 eggs" in the oval. (*Pause for students to write.*)

We underlined the two things (<u>cookies</u> and <u>eggs</u>) in the problem that formed a ratio and wrote their names in the two columns in the diagram, circled the numbers, wrote the numbers and labels for what we know, and wrote a "?" for what must be solved in the vary diagram. Check off the second and third boxes under Step 2. Now let's look at the diagram and read what it says.

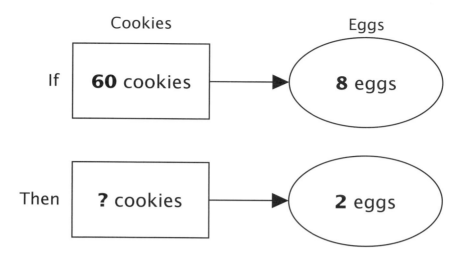

(*Point to relevant parts of the diagram as you explain.*) If "60 cookies use 8 eggs," then "? cookies use 2 eggs," or how many almond cookies can she make with 2 eggs? What must you solve for in this problem?

Students: Number of cookies that can be made with 2 eggs in the second pair of association in the diagram.

Teacher: Great! Now for Step 3: Plan to solve the problem. (*Point to first box under Step 3 on checklist.*) To plan to solve the problem, we need to translate the information in the diagram into a math equation:

$$\frac{60 \text{ cookies}}{? \text{ cookies}} = \frac{8 \text{ eggs}}{2 \text{ eggs}}$$

OR

$$\frac{60}{?} = \frac{8}{2}$$

Let's check off the box under Step 3 on the checklist. Now we are ready for Step 4: Solve the problem. (*Point to first box under Step 4 on checklist.*) We need to solve for the unknown or question mark in the equation. I will solve for the "?" by applying the cross-multiplication method:

$$8 \times ? = 60 \times 2$$
$$? = (60 \times 2) \div 8$$
$$? = 15$$

Now check off the first box under Step 4 on the checklist. (*Point to second box under Step 4 on checklist.*) Next write 15 for the "?" in the diagram and write the complete answer on the answer line. The complete answer is the number and the label. What is the complete answer to this vary problem?

Students: 15 cookies.

Teacher: Good. Write 15 cookies on the answer line. (*Pause for students to write the answer.*) Let's check off the second box under Step 4 on the checklist. (*Point to third box under Step 4 on checklist.*) We are now ready to check the answer. Does "15 cookies" seem right?

Let's see ... if 2 eggs make 15 cookies, 8 eggs will make 4 times of 15, that is 4 × 15 = 60. So our answer (i.e., 15 cookies) seems right. We can also check by redoing the cross-multiplication using all the numbers. That is, 60 × 2 = 15 × 8, or 60 = 60. So my calculation is correct. (*Check off the third box under Step 4.*)

(*Guide students to write the explanation for solving the problem on their worksheet; see Vary Reference Guide 1.*)

Let's review this vary problem. This problem tells about an association (i.e., ratio) between the number of cookies that can be made and the number of eggs needed to make them. The *if–then* statements make up two pairs of associations or two ratios, and the two ratios are equivalent. What is this problem called? How do you know?

Students: Vary, because the problem tells about a ratio between cookies and eggs. The *if–then* statement makes up two pairs of associations or ratios that are equivalent.

Vary Problem 3

Teacher: (*Use the script as a guideline for solving Vary Problem 3, and facilitate problem solving by having frequent student–teacher exchanges. Display Overhead Modeling page of Vary Problem 3. Have students look at Vary Problem 3.*)
Touch Vary Problem 3. What's the first step?

Students: Find the problem type.

Teacher: Right. (*Point to first check box on Vary Problem-Solving Checklist.*) To find the problem type, what must you do?

Students: Read the problem and retell it in own words.

Teacher: Good. Read the problem aloud.

Students: "A farmer has 444 eggs to pack in cartons. If each egg carton holds 12 eggs, how many cartons will the farmer need?"

Teacher: You read the problem. What must you do next?

Students: Retell it using own words.

Teacher: Yes. Now retell the problem in your own words to help you understand it. (*Call on students to retell the problem. When they retell* the problem, remind students to tell what they know in the problem and what they are asked to find out.)

Students: This story tells about an association or a ratio between the number of eggs and the number of cartons needed (to pack eggs). I know the farmer has 444 eggs to pack in cartons. I also know that each egg carton holds 12 eggs. I do not know how many cartons the farmer will need to pack 444 eggs.

Teacher: Check off the first box under Step 1 on the checklist. (*Point to second check box under Step 1.*) What kind of problem is this and how do you know?

Students: A vary problem, because it tells about a ratio or rate between two things and there is an *if–then* statement.

Teacher: Does this problem tell about a rate or ratio between two things?

Students: The problem tells about an association or a ratio between the number of eggs and the number of cartons.

Teacher: Is there an *if–then* statement that makes up two pairs of associations? How do you know?

Students: Yes, the second sentence in this problem is the *if* statement that tells about a specific ratio between the numbers of cartons and eggs (i.e., "If each egg carton holds 12 eggs"). Also, the second pair of association is about how many cartons are needed to pack 444 eggs. So there are two pairs of association between cartons and eggs. Therefore, this problem is a vary problem.

Teacher: Great! We can restate the problem as follows: "*If* each egg carton holds 12 eggs, *then* how many cartons are needed to pack 444 eggs?" Let's check off the second box under Step 1. Now we are ready for Step 2: Organize the information in the problem using the vary diagram. (*Display Vary Problem diagram poster.*)

Vary Problem

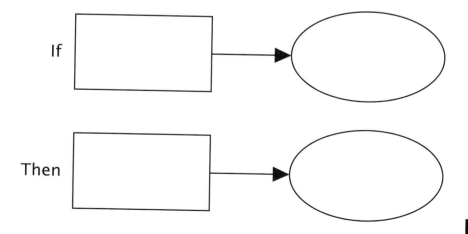

(*Point to first box under Step 2.*) To organize the information in a vary problem, what must you do first?

Students: We first underline the two things that form a specific ratio and write their names in the diagram.

Teacher: What does this problem talk about? How do you know?

Students: This problem talks about a specific ratio between cartons and eggs, because both the *if* statement and the question (i.e., the *then* statement) talk about the number of cartons and the number of eggs that can be packed in the cartons.

Teacher: Good! Underline <u>eggs</u> and <u>cartons</u> in the problem and write "Cartons" for the first column and "Eggs" for the second column in the diagram. (*Remind students that it is okay to write either "eggs" or "cartons" in the first column; however, it is important to align the numbers for cartons and eggs in the correct columns. Pause for students to complete.*) Check off the first box under Step 2 on the checklist. (*Point to second box under Step 2.*)

Now read each sentence to circle numbers for each of the two pairs of associations and write numbers and labels in the diagram. Typically, we find the specific ratio given in the problem and write the numbers for the *if* statement in the diagram. Does the first sentence in the problem tell about a specific ratio between cartons and eggs?

Students: No. It just says, "A farmer has 444 eggs to pack in cartons."

Teacher: Does the second sentence tell about a specific ratio?

Students: Yes. The second sentence says "each egg carton holds 12 eggs."

Teacher: So what numbers do you circle and write for the *if* statement in the diagram?

Students: We circle "each" and write "1 carton" in the box for the "Carton" column and circle "12" and write "12 eggs" in the oval for the "Eggs" column.

Teacher: Great! (*Pause for students to complete.*) Now you need to circle and write numbers and labels for the *then* statement in the diagram. Do you know the number for the cartons in the *then* part of the diagram?

Students: No. This is what we are asked to solve. The problem asks how many cartons the farmer will need.

Teacher: Correct. We do not know how many cartons are needed to pack 444 eggs. Let's write a "?" in the box for the "Cartons" column. (*Pause for students to complete.*) Do you know the number for the "Eggs" column in the *then* part of the diagram?

Students: Yes. The first sentence in the problem says, "A farmer has 444 eggs to pack in cartons."

Teacher: Great! Circle "444" and write "444 eggs" in the oval. (*Pause for students to write.*)

We underlined the two things (cartons and eggs) in the problem that formed a ratio and wrote their names in the two columns in the diagram, circled numbers, wrote numbers and labels for what we know, and wrote a "?" for what must be solved in the vary diagram. Check off the second and third boxes under Step 2. Now let's look at the diagram and read what it says.

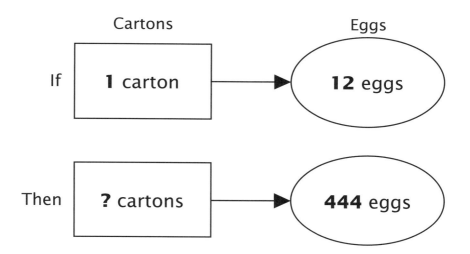

(*Point to relevant parts of the diagram as you explain.*) If "1 carton holds 12 eggs," then "? cartons hold 444 eggs," or how many cartons are needed to hold 444 eggs? What must you solve for in this problem?

Students: Number of cartons to hold 444 eggs.

Teacher: Great! Now for Step 3: Plan to solve the problem. (*Point to first box under Step 3 on checklist.*) To plan to solve the problem, we need to translate the information in the diagram into a math equation. What is the math equation?

Students:

$$\frac{1 \text{ carton}}{? \text{ cartons}} = \frac{12 \text{ eggs}}{444 \text{ eggs}}$$

OR

$$\frac{1}{?} = \frac{12}{444}$$

Teacher: Check off the box under Step 3 on the checklist. Now we are ready for Step 4: Solve the problem. (*Point to first box under Step 4* *on checklist.*) We need to solve for the unknown or question mark in the equation. What method can you use to solve for the "?" in the equation?

Students: Cross-multiplication.

Teacher: Good. Now solve the problem.

Students: 12 × ? = 1 × 444
? = 444/12
? = 37

Teacher: Now check off the first box under Step 4 on the checklist. (*Point to second box under Step 4 on checklist.*) What is the complete answer to this problem?

Students: 37 cartons.

Teacher: Good. Now write 37 for the "?" in the diagram and write the complete answer on the answer line. (*Pause for students to write the answer.*) Let's check off the second box under Step 4 on the checklist. (*Point to third box under Step 4 on checklist.*) We are now ready to check the answer. Does "37 cartons" seem right?

Students: If one carton holds 12 eggs, then 37 cartons will hold 37 times of 12, that is 37 × 12 = 444. So the answer 37 cartons seems right.

Teacher: We can also check by redoing the cross-multiplication using all the numbers. What do you get when you redo the cross-multiplication?

Students: 37 × 12 = 1 × 444, or 444 = 444.

Teacher: So your calculation is correct. (*Check off third box under Step 4.*) Let's review this vary problem. This problem tells about an association (i.e., ratio) between number of cartons and numbers of eggs. The *if–then* statements make up two pairs of associations or two ratios, and the two ratios are equivalent. What is this problem called? How do you know?

Students: Vary, because the problem tells about a ratio between cartons and eggs and the *if–then* statement makes up two equivalent ratios.

Teacher: Great job working hard. Tomorrow we will practice more vary problems.

Lesson 8: Problem Solution

Materials Needed

Answer Sheet for Paired-Learning Tasks	Vary Answer Sheet 1
Checklists	Word Problem–Solving Steps (FOPS) poster
	Vary Problem–Solving Checklist (laminated copies for students)
Diagram	Vary Problem diagram
Overhead Modeling	Vary Problems 4 and 5
Student Pages	Vary Problems 4 and 5
	Vary Worksheet 1

Vary Problem 4

Teacher: [*Display Word Problem–Solving Steps (FOPS) poster. Ask students to read each step on the poster that they will use to solve multiplication and division word problems. Display Overhead Modeling page of Vary Problem 4. Have students look at Vary Problem 4.*]

 Touch Problem 4. (*Point to Vary Problem–Solving Checklist.*) Remember, we will use this checklist that has the same four steps (FOPS) to help us solve vary problems. What's the first step?

Students: Find the problem type.

Teacher: Right. (*Point to first check box on checklist.*) To find the problem type, what must you do?

Students: Read the problem and retell it in own words.

Teacher: Good. I will read the problem aloud.
 "Ms. Hanna used 12 strawberries to make 5 cups of fruit cocktail. If she needs to make 20 cups of fruit cocktail for a party, how many strawberries does she need?"
 I read the problem. What must you do next?

Students: Retell the problem using own words.

Teacher: Yes. Now retell the problem in your own words to help you understand it. (*Call on students to retell the problem. When they retell the problem, remind students to tell what they know in the problem and what they are asked to find out.*)

Students: I know Ms. Hanna used 12 strawberries to make 5 cups of fruit cocktail. I also know she needs to make 20 cups of fruit cocktail. I do not know how many strawberries she needs to make 20 cups of fruit cocktail.

Teacher: Good. Check off the first box under Step 1 on the checklist. (*Point to second check box under Step 1.*) What kind of problem is this? How do you know?

Students: The problem tells about an association or a ratio between the number of strawberries and the cups of fruit cocktail. The first sentence in this problem is the *if* statement that tells about a specific ratio between the number of strawberries and the cups of fruit cocktail (i.e., 12 strawberries make 5 cups of fruit cocktail). Also, the second pair of association is about how many strawberries are needed to make 20 cups of fruit cocktail. So there are two pairs of association between strawberries and cups of fruit cocktail. Therefore, this problem is a vary problem.

Teacher: Great! We can restate the problem as follows: "If Ms. Hanna used 12 strawberries to make 5 cups of fruit cocktail, then how many strawberries are needed to make 20 cups of fruit cocktail?" Let's check off the second box under Step 1. Now we are ready for Step 2: Organize the information in the problem using the vary diagram. (*Display Vary Problem diagram poster.*)

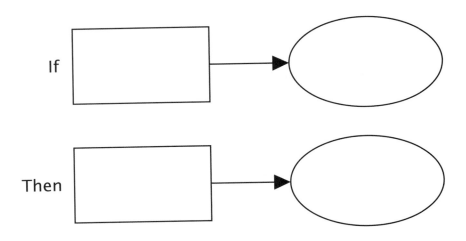

(*Point to first box under Step 2 of checklist.*) To organize the information in a vary problem, what must you do first?

Students: We first underline the two things that form a specific ratio and write their names in the diagram.

Teacher: What does this problem talk about? How do you know?

Students: This problem talks about a specific ratio between strawberries and cups of fruit cocktail, because both the *if* statement and the question (i.e., the *then* statement) talk about the number of strawberries and cups of fruit cocktail.

Teacher: Good! Underline <u>strawberries</u> and <u>cups of fruit cocktail</u> in the problem and write "Strawberries" for the first column and "Cups of fruit cocktail" for the second column in the diagram. (*Remind students that it is okay to write either "strawberries" or "cups of fruit cocktail" in the first column; however, it is important to align the numbers for strawberries and cups of fruit cocktail in the correct columns. Pause for students to complete.*)

Check off the first box under Step 2 on the checklist. (*Point to second box under Step 2.*) Now read each sentence to circle numbers for each of the two pairs of associations and write numbers and labels in the diagram. Typically, we first find the specific ratio given in the problem and write the numbers for the *if* statement in the diagram. Does the first sentence in the problem tell about a specific ratio between strawberries and cups of fruit cocktail?

Students: Yes, the first sentence says "Ms. Hanna used 12 strawberries to make 5 cups of fruit cocktail."

Teacher: So what numbers do you circle and write for the *if* statement in the diagram?

Students: We circle "12" and write "12 strawberries" in the box for the "Strawberries" column and circle "5" and write "5 cups" in the oval for the "Cups of fruit cocktail" column.

Teacher: Great! (*Pause for students to complete.*) Now you need to circle and write numbers and labels for the *then* statement in the diagram. From the question (i.e., the *then* statement), do you know the number of "strawberries" to write in the *then* part of the diagram?

Students: No. This is what we are asked to solve. The problem asks how many strawberries are needed to make 20 cups of fruit cocktail.

Teacher: Correct. We do not know how many strawberries are needed to make 20 cups of fruit cocktail. Let's write "? strawberries" in the box for the "Strawberries" column. (*Pause for students to complete.*) Do you know the number for the "Cups of fruit cocktail" column in the *then* part of the diagram?

Students: Yes, 20 cups of fruit cocktail.

Teacher: Great! Circle "20" and write "20 cups" in the oval. (*Pause for students to write.*) We underlined the two things ("strawberries" and "cups of fruit cocktail") in the problem that formed a ratio and wrote their names for the two columns in the diagram, circled numbers, wrote numbers and labels for what we know, and wrote a "?" for what must be solved in the vary diagram. Check off the second and third boxes under Step 2. Now let's look at the diagram and read what it says.

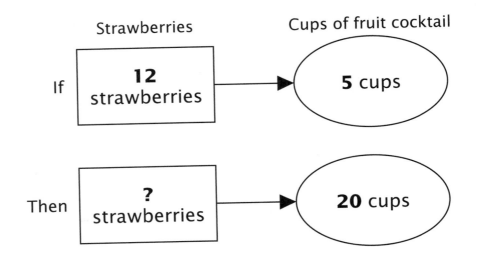

(*Point to relevant parts of the diagram as you explain.*) If "12 strawberries make 5 cups of fruit cocktail," then "? strawberries make 20 cups of fruit cocktail," or how many strawberries are needed to make 20 cups of fruit cocktail. What must you solve for in this problem?

Students: Number of strawberries to make 20 cups of fruit cocktail.

Teacher: Great! Now for Step 3: Plan to solve the problem. (*Point to first box under Step 3 on checklist.*) To plan to solve the problem, we need to translate the information in the diagram into a math equation. What is the math equation?

Students:

$$\frac{12 \text{ strawberries}}{? \text{ strawberries}} = \frac{5 \text{ cups}}{20 \text{ cups}}$$

OR

$$\frac{12}{?} = \frac{5}{20}$$

Teacher: Check off the first box under Step 3 on the checklist. Now we are ready for Step 4: Solve the problem. (*Point to first box under Step 4 on checklist.*) We need to solve for the unknown or question mark in the equation. What method can you use to solve for the "?" in the equation?

Students: Cross-multiplication.

Teacher: Good, now solve the problem.

Students: ? × 5 = 12 × 20
? = 240/5
? = 48

Teacher: Now check off the first box under Step 4 on the checklist. (*Point to second box under Step 4 on checklist.*) What is the complete answer to this problem?

Students: 48 strawberries.

Teacher: Good. Now write 48 for the "?" in the diagram and write the complete answer on the answer line. (*Pause for students to write the answer.*) Let's check off the second box under Step 4 on the checklist. (*Point to third box under Step 4.*) We are now ready to check the answer. Does "48 strawberries" seem right?

Students: If 12 strawberries make 5 cups of fruit cocktail, then 48 strawberries will make 4 times of 5 cups, that is 4 × 5 = 20. So the answer 20 cups seems right.

Teacher: We can also check by redoing the cross-multiplication using all the numbers. What do you get when you redo the cross-multiplication?

Students: 48 × 5 = 12 × 20, or 240 = 240.

Teacher: So your calculation is correct. (*Check off the third box under Step 4.*) Let's review this vary problem. This problem tells about an association (i.e., ratio) between the number of strawberries and the cups of fruit cocktail. The *if–then* statements make up two pairs of associations or two ratios, and the two ratios are equivalent. What is this problem called? How do you know?

Students: Vary, because the problem tells about a ratio between strawberries and cups of fruit cocktail and the *if–then* statement makes up two equivalent ratios.

Vary Problem 5

Teacher: (*Display Overhead Modeling page of Vary Problem 5. Have students look at Vary Problem 5.*)
(*Point to first check box on Vary Problem-Solving Checklist.*) To find the problem type, what must you do?

Students: Read the problem and retell it in own words.

Teacher: Good. I will read the problem aloud.
"If 3 centimeters on the map represent 12 miles on the road, how many miles will 5 centimeters represent?"
I read the problem. Now you retell the problem using your own words.

Students: I know 3 centimeters (cm) on the map represent 12 miles on the road. I do not know the number of miles that 5 cm represent.

Teacher: Good. Check off the first box under Step 1 on the checklist. (*Point to second check box under Step 1.*) What kind of problem is this? How do you know?

Students: It is a vary problem, because it tells about an association or a ratio between the number of centimeters on the map and the number of miles on the road. The first sentence in this problem is the *if* statement that tells about a specific ratio between centimeters (3) on the map and miles (12) on the road. Also, the second pair of association is about how many miles 5 cm will represent. So there are two pairs of associations between centimeters on the map and miles on the road. Therefore, this problem is a vary problem.

Teacher: Great! Let's check off the second box under Step 1. Now we are ready for Step 2: Organize the information using the vary diagram. (*Display Vary Problem diagram poster.*)

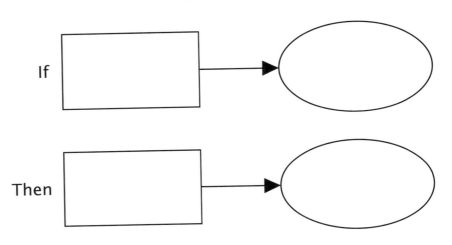

(*Point to first box under Step 2 of checklist.*) To organize the information in a vary problem, what must you do first?

Students: We first underline the two things that form a specific ratio and write their names in the diagram.

Teacher: What does this problem talk about? How do you know?

Students: This problem talks about a specific ratio between centimeters on the map and miles on the road, because both the *if* statement and the question talk about the number of centimeters on the map and the number of miles on the road.

Teacher: Good! Underline <u>centimeters</u> and <u>miles</u> in the problem, and write "Centimeters" for the first column and "Miles" for the second column in the diagram. (*Remind students that it is okay to write either "centimeters" or "miles" in the first column; however, it is important to align* *the numbers for centimeters and miles in the correct columns. Pause for students to complete.*)

Check off the first box under Step 2 on the checklist. (*Point to second box under Step 2.*) Now read each sentence to circle numbers for each of the two pairs of associations and write numbers and labels in the diagram. Typically, we find the specific ratio given in the problem and write the numbers for the *if* statement in the diagram. Does the first sentence in the problem tell about a specific ratio between centimeters on the map and miles on the road?

Students: Yes, the first sentence says "3 centimeters on the map represent 12 miles on the road."

Teacher: So what numbers do you circle and write for the *if* statement in the diagram?

Students: We circle "3" and write "3 cm" in the box for the "Centimeters" column and circle "12" and write "12 miles" in the oval for the "Miles" column.

Teacher: Great! (*Pause for students to complete.*) Now you need to circle and write numbers and labels for the *then* statement in the diagram. Do you know the number for centimeters in the *then* part of the diagram?

Students: Yes. The problem asks for the number of miles that 5 centimeters represent. Therefore, we know the number for centimeters in the *then* part of the diagram.

Teacher: Great! Circle "5" and write "5 cm" in the box. (*Pause for students to write.*) Do you know the number for the "Miles" column in the *then* part of the diagram?

Students: No. This is what we are asked to solve.

Teacher: Correct. We do not know how many miles 5 cm on the map will represent. Let's write "? miles" in the oval for the "Miles" column. (*Pause for students to complete.*) We underlined the two things (<u>centimeters</u> and <u>miles</u>) in the problem that formed a ratio and wrote their names in the two columns in the diagram, circled numbers, wrote numbers and labels for what we know, and wrote a "?" for what must be solved in the vary diagram. Check off the second and third boxes under Step 2. Now let's look at the diagram and read what it says.

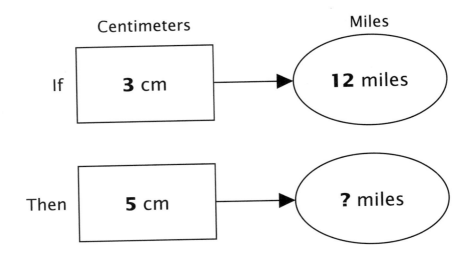

(*Point to relevant parts of the diagram as you explain.*) If 3 cm on the map represents 12 miles on the road, then 5 cm will represent how many miles on the road, or how many miles will 5 cm represent. What must you solve for in this problem?

Students: Number of miles that 5 cm will represent.

Teacher: Great! Now for Step 3: Plan to solve the problem. (*Point to first box under Step 3 on checklist.*) To plan to solve the problem, we need to translate the information in the diagram into a math equation. What is the math equation?

Students:

$$\frac{3 \text{ cm}}{5 \text{ cm}} = \frac{12 \text{ miles}}{? \text{ miles}}$$

OR

$$\frac{3}{5} = \frac{12}{?}$$

Teacher: Check off the first box under Step 3 on the checklist. Now we are ready for Step 4: Solve the problem. (*Point to first box under Step 4 on checklist.*) We need to solve for the unknown or question mark in the equation. What method can you use to solve for the "?" in the equation?

Students: Cross-multiplication.

Teacher: Good, now solve the problem.

Students: 3 × ? = 5 × 12
  ? = (5 × 12) ÷ 3
  ? = 20

Teacher: Now check off the first box under Step 4 on the checklist. (*Point to second box under Step 4.*) What is the complete answer to this problem?

Students: 20 miles.

Teacher: Good. Now write 20 for the "?" in the diagram and write the complete answer on the answer line. (*Pause for students to write the answer.*) Let's check off the second box under Step 4 on the checklist. (*Point to third box under Step 4.*) We are now ready to check the answer. Does 20 miles seem right?

Students: If 3 cm on the map represents 12 miles on the road, then 5 cm will represent a little less than 2 times of 12, that is, less than 24. So the answer 20 miles seems right.

Teacher: We can also check by redoing the cross-multiplication using all the numbers. What do you get when you redo the cross-multiplication?

Students: 3 × 20 = 5 × 12, or 60 = 60.

Teacher: So your calculation is correct. (*Check off third box under Step 4.*) Let's review this vary problem. This problem tells about an association (i.e., ratio) between the number of centimeters on the map and the number of miles on the road. The *if–then* statements make up two pairs of associations or two ratios, and the two ratios are equivalent. What is this problem called? How do you know?

Students: Vary, because the problem tells about a ratio between centimeters on the map and miles on the road and the *if–then* statement makes up two equivalent ratios. (*Pass out Vary Worksheet 1.*)

Teacher: Now I want you to do Vary Worksheet 1, Problem 1 with your partner.
  (*Ask students to think, plan, and share with partners to solve Vary Problem 1 on the worksheet; see Guide to Paired Learning in the Introduction.*)

Vary Worksheet 1, Problem 1: "If Rich can save $48 in 5 months, how many months will it take for Rich to save $144?"

(*Monitor students as they work. Have students check their answers using Vary Answer Sheet 1. Make sure the diagram is labeled correctly, the math equation is written and worked out correctly, the written explanation is complete, and the complete answer is written on the answer line; see below.*)

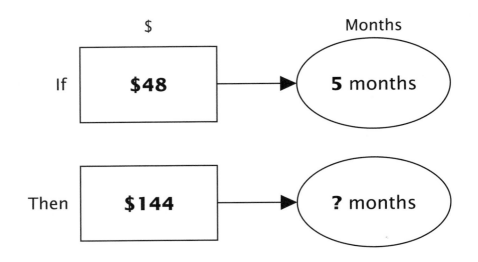

Answer: 15 months

Teacher: Now I want you to do the next two problems on your own. Remember to use the four steps to solve these problems.

Vary Worksheet 1, Problem 2: "Ben can eat 7 hot dogs in 2 minutes. How many hot dogs can he eat in 6 minutes?"

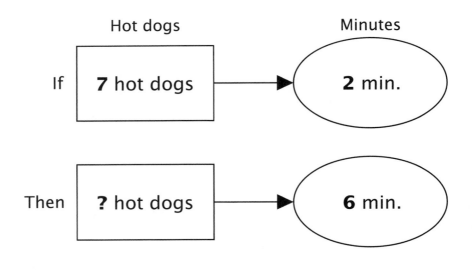

Answer: 21 hot dogs

Vary Worksheet 1, Problem 3: "There are 4 chocolate bunnies in each candy box. If you bought 12 boxes, how many chocolate bunnies would you have?"

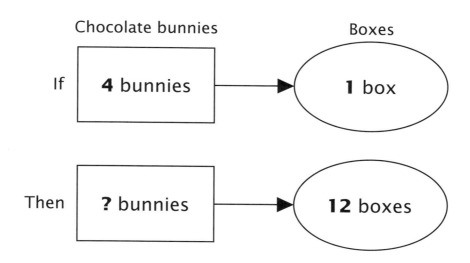

Answer: 48 chocolate bunnies

Teacher: (*Monitor students as they work. After about 10 minutes, go over the answers. Make sure the diagrams are labeled correctly, the math equations are worked out correctly, and the complete answers are written on the answer lines.*)

Great job working hard. Tomorrow we will practice more vary problems.

Lesson 9: Problem Solution

Materials Needs

Answer Sheet for Paired-Learning Tasks	Lesson 9: Vary Answer Sheet 2
Checklists	Word Problem–Solving Steps (FOPS) poster
	Vary Problem–Solving Checklist (laminated copies for students)
Diagram	Vary Problem diagram poster
Overhead Modeling	Lesson 9: Vary Problem 1
Student Pages	Lesson 9: Vary Worksheet 2

Teacher: (*Pass out student copies of Vary Worksheet 2. Display Overhead Modeling page of Vary Problem 1.*)

Follow along as I read this problem. (*Use guided practice to have students complete Vary Worksheet 2, Problem 1. Read the problem aloud.*)

"Use data from the table below to solve Vary Worksheet 2, Problem 1."

Items on Sale

Payless Ice Cream	$2.50 for 2 half-gallon boxes
Jay's Potato Chips	$3.00 for 2 bags
Perdue Chicken Quarters	$1.25 for 5-pound package

Vary Worksheet 2, Problem 1: "Above is a list of items on sale at Payless Grocery Store. If you plan to buy 18 bags of Jay's Potato Chips, how much do you have to pay?" (*Use the four steps to solve a vary problem.*)

(*Assist students, if necessary, to identify needed information from the table provided. Because the problem asks for the cost of 18 bags of Jay's Potato Chips, only information regarding Jay's Potato Chips is relevant. Assist students in retelling the problem so that the problem is reconstructed as follows: The store wants $3.00 for 2 bags of Jay's Potato Chips. If you plan to buy 18 bags of the potato chips, how much do you have to pay?*)

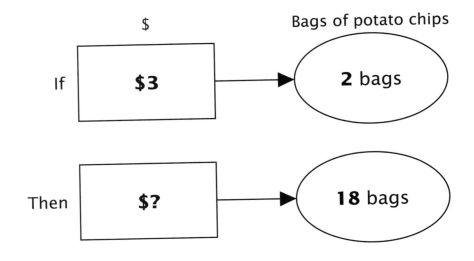

Answer: $27

Teacher: Now I want you to do the next problem with your partner.
(Ask students to think, plan, and share with partners to solve Problem 2 on the worksheet; see Guide to Paired Learning in the Introduction.)

Vary Worksheet 2, Problem 2: "On Saturday, all the seats in Stone Theater were taken by a total of 384 people. If there are 16 rows of seats in the theater, how many seats are there in each row?"

(Use the four steps to solve this vary problem. Monitor students as they work. Have students check their answers using Vary Answer Sheet 2. Make sure the diagram is labeled correctly, the math equation is written and worked out correctly, the written explanation is complete, and the complete answer is written on the answer line; see below.)

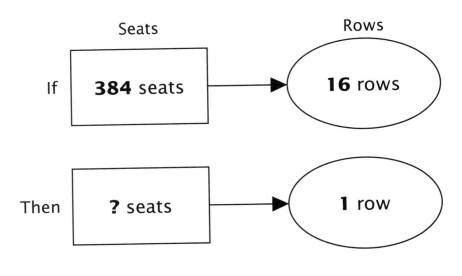

Answer: 24 seats

Teacher: Now I want you to do the next four problems on your own. Remember to use the four steps to solve the problems on this worksheet.

"Use the recipe below to solve Vary Worksheet 2, Problem 3."

Chinese Dumpling Recipe (for 30 dumplings)

5 cups of all-purpose flour

1 lb. ground pork meat

1/3 cup vegetable oil

3 lb. minced vegetables

1 tablespoon cooking wine

1/4 teaspoon minced fresh ginger

Vary Worksheet 2, Problem 3: "If Ms. Lee needs to make 120 dumplings, how many cups of flour does she need?"

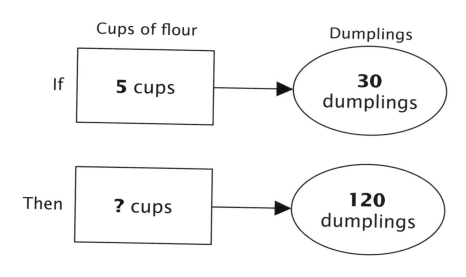

Answer: 20 cups of flour

Vary Worksheet 2, Problem 4: "Each notebook has 70 pages in it. If you bought 5 notebooks, how many pages would you have in all?"

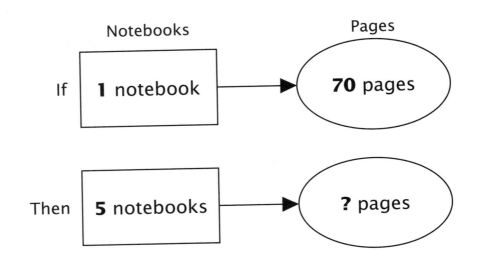

Answer: 350 pages

Vary Worksheet 2, Problem 5: "A total of 48 slices of cake are ready to serve the teachers and students sitting around several tables. If 8 slices of cake will be served at each table, how many tables can be fully served?"

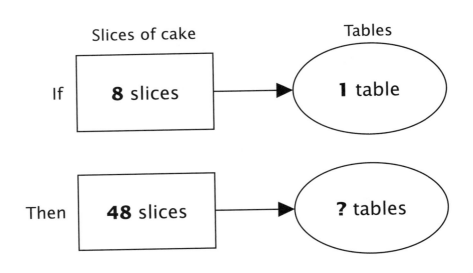

Answer: 6 tables

Vary Worksheet 2, Problem 6: "Julia spent 3 hours making 12 Christmas ornaments. How many hours will it take for her to make 36 ornaments?"

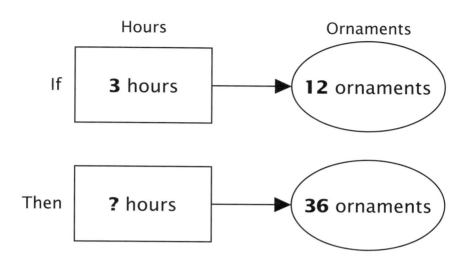

Answer: 9 hours

Teacher: (*Monitor students as they work. After about 15 to 20 minutes, go over the answers. Make sure the diagrams are labeled correctly, the math equations are written and worked out correctly, and the complete answers are written on the answer line.*)

Great job working hard. Tomorrow you will solve more vary problems using your own diagrams.

Lesson 10: Problem Solution

Materials Needed

Checklists	Word Problem–Solving Steps (FOPS) checklist
	Vary Problem–Solving Checklist (laminated copies for students)
Overhead Modeling	Lesson 10: Vary Problem 1
Reference Guide	Lesson 10: Vary Reference Guide 2
Student Pages	Lesson 10: Vary Worksheet 3

Teacher: (*Pass out Vary Worksheet 3. Display Overhead Modeling page for Vary Problem 1.*)

You learned to solve vary problems using diagrams. Now we will solve the problems on this worksheet using your own diagrams. This worksheet does not have diagrams. Remember to use the four steps (FOPS) to solve problems on the worksheet. (*Discuss how students can generate a diagram that is more efficient than the one they used, and have them practice solving the problems using the diagram they generate. Also, encourage them to use the Vary Problem-Solving Checklist only as needed. Use guided practice to complete Problems 1 and 2 using own diagrams; see below.*)

"Use the data below to solve Vary Worksheet 3, Problem 1."

Number of books	10	...	...	60
Number of wrapping papers needed	?	...	...	12

Vary Worksheet 3, Problem 1: "Students are required to wrap their books within the first week of the new school semester. If Joy needs to wrap 10 books, how many wrapping papers does she need?" (*See Vary Reference Guide 2.*)

$$\text{If} \quad \frac{60 \text{ books}}{10 \text{ books}} = \frac{12 \text{ papers}}{? \text{ papers}}$$
$$\text{Then}$$

Answer: 2 wrapping papers

Vary Worksheet 3, Problem 2: "The cost of 3 Sneaker chocolate bars is 2 dollars. If you have 6 dollars, how many Sneaker candy bars can you buy?"

$$\text{If} \quad \frac{3 \text{ bars}}{? \text{ bars}} = \frac{\$2}{\$6} \quad \text{Then}$$

Answer: 9 bars

Teacher: Now I want you to do the next three problems on your own. (*Have students write the explanation for at least one of the three problems.*) Remember to use the four steps to solve Problems 3 through 5 on the worksheet.

Vary Worksheet 3, Problem 3: "The art class is going on a field trip. Only 4 buses are available for the trip. If each bus will seat 26 students, how many students in all can be seated in the buses for the field trip?"

$$\text{If} \quad \frac{1 \text{ bus}}{4 \text{ buses}} = \frac{26 \text{ students}}{? \text{ students}} \quad \text{Then}$$

Answer: 104 students

Vary Worksheet 3, Problem 4: "Ann planted 4 rows of tomato plants in her vegetable garden yesterday. There were 6 tomato plants in each row. How many tomato plants in all were in the garden?"

$$\text{If} \quad \frac{6 \text{ plants}}{? \text{ plants}} = \frac{1 \text{ row}}{4 \text{ rows}} \quad \text{Then}$$

Answer: 24 plants

Vary Worksheet 3, Problem 5: "An automobile used 8 gallons of gasoline to travel 120 miles. At that rate, how many gallons will be used to travel 360 miles?"

$$\text{If} \quad \frac{8 \text{ gallons}}{? \text{ gallons}} = \frac{120 \text{ miles}}{360 \text{ miles}}$$
$$\text{Then}$$

Answer: 24 gallons

Teacher: You learned to solve vary word problems using your own diagrams. Next you will learn to solve both MC and vary problems when they are mixed.

Unit 3

Multiplicative Compare and Vary Problem Review

Lesson 11: Multiplicative Compare and Vary Review

Materials Needed

Checklists	MC Problem–Solving Checklist and Vary Problem–Solving Checklist posters
	MC Problem–Solving Checklist and Vary Problem–Solving Checklist (laminated copies for students)
Diagram	Multiplicative Compare Problem and Vary Problem diagram posters
Overhead Modeling	Lesson 11: Review Problems 1 and 2
Student Pages	Lesson 11: Review Problems 1 and 2

Teacher: *(Display posters of Multiplicative Compare Problem and Vary Problem diagram posters and the two checklist posters in the classroom in a suitable place. Pass out Review Problems 1 and 2.)*
Today we are going to review MC and vary problems. You will practice these problems to make sure you remember them. Let's review the MC problem. What have you learned about an MC problem?

Students: An MC problem compares one quantity (the compared) to the other (the referent), and the compare words such as *3 times as many as* or *1/3 as many as* in the problem indicate a multiple or partial relation between the compared and the referent.

Teacher: What have you learned about a vary problem?

Students: A vary problem tells about a ratio between two things and the *if-then* statement makes up two equivalent ratios.

Teacher: Great! What are the four steps (FOPS) to solving all addition and subtraction or multiplication and division problems?

Students: Find the problem type; Organize information in the problem using diagrams; Plan to solve the problem; and Solve the problem.

Teacher: Good job. (*Point to MC Problem and Vary Problem diagram posters.*) You learned to use these diagrams to organize information to solve MC and vary problems. Later you used your own diagrams to organize information and solve these two different types of problems. Today you will learn to identify each type of problem and solve it using the problem-solving steps you learned earlier.

Review Problem 1

Teacher: Let's look at Review Problem 1. (*Display Overhead Modeling page of Review Problem 1. Have students look at Review Problem 1.*) "Use the recipe below to solve Review Problem 1."

Chocolate Cupcake Recipe (for 18 cupcakes)

1 pack chocolate cake mix

3 whole eggs

1/3 cup vegetable oil

1-1/2 cups water

"If you are planning to make 126 cupcakes for your school's teacher appreciation party, how many eggs do you need?"

Teacher: Touch Review Problem 1. (*Guide students to identify the problem type and solve the problem.*) What's the first step in FOPS?

Students: Find the problem type.

Teacher: That's right. To find the problem type, what must you do first?

Students: Read the problem and retell it in your own words.

Teacher: Good. (*Read the problem and call on a student to retell the problem.*) Remember, when you retell a problem, you use your own words and tell what you know in the problem and what you are asked to find out.

Students: We know from the recipe that 3 eggs are needed to make 18 cupcakes. We need to find out how many eggs are needed to make 126 cupcakes.

Teacher: Great! You read and retold the problem. Now you need to ask yourself if this is an MC or a vary problem. What kind of problem is this? How do you know?

Students: It is a vary problem, because it tells about a ratio between the number of eggs and the number of cupcakes. One pair of association or ratio (eggs to cupcakes) is given in the recipe (i.e., 3 eggs make 18 cupcakes), which is the *if* statement. We need to find out the number of eggs needed to make 126 cupcakes, or the *then* part of the statement.

Teacher: Great, we can restate the problem as follows: "*If* 3 eggs make 18 cupcakes, *then* how many eggs are needed to make 126 cupcakes?"

Now you are ready for Step 2: Organize the information in the problem using the vary diagram. To organize the information in your vary diagram, what must you do first?

Students: We first underline the two things that form a specific ratio and write their names in the diagram.

Teacher: What does this problem talk about? How do you know?

Students: This problem talks about a specific ratio between eggs and cupcakes, because both the *if* statement and the question (i.e., the *then* statement) talk about the numbers of eggs and cupcakes.

Teacher: Excellent! Write their names in your diagram. (*Pause for students to complete.*)

Now read the problem and find the numbers for each of the two pairs of associations, and write the numbers and labels in the diagram. Remember, we first find the specific ratio given in the problem and write the numbers for the *if* statement in the diagram. What is the specific ratio of eggs to cupcakes given in the problem?

Students: It is 3 eggs to 18 cupcakes.

Teacher: Great! Write the numbers and labels in your diagram for the *if* statement. (*Pause for students to complete.*) Now circle and write numbers and labels for the *then* statement in the diagram. Also, remember to write a question mark for the unknown information. (*Pause for students to figure out and check their work using the following completely mapped diagram.*)

$$\begin{array}{c} \text{If} \\ \text{Then} \end{array} \quad \dfrac{3 \text{ eggs}}{? \text{ eggs}} = \dfrac{18 \text{ cupcakes}}{126 \text{ cupcakes}}$$

Teacher: Now look at the diagram and read what it says. (*Pause for students to read.*) What must you solve for in this problem?

Students: The number of eggs in the *then* part of the diagram.

Teacher: Good! You completed Step 2 using FOPS. What is the next step?

Students: Plan to solve the problem.

Teacher: Good. To solve this vary problem, all you need to do is translate your diagram into a math equation. What is the math equation?

Students:

$$\dfrac{3}{?} = \dfrac{18}{126}$$

Teacher: Excellent. You completed Step 3 using FOPS. What is Step 4?

Students: Solve the problem.

Teacher: Good. To solve for the question mark in the equation, we can use cross-multiplication. Now use cross-multiplication to solve the problem.

Students: $3 \times 126 = ? \times 18$
$? = (3 \times 126) \div 18$
$? = 21$

Teacher: What is the complete answer to this vary problem?

Students: 21 eggs.

Teacher: Good. Write "21 eggs" on the answer line and check your answer. (*Pause for students to write the answer.*) Does "21 eggs" seem right? How do you know?

Students: Yes. If 3 eggs make 18 cupcakes, then 21 eggs make 7 times of 18 or 126 eggs. So the answer "21 eggs" seems right.

Teacher: Excellent work! Let's solve the next problem on your worksheet.

Review Problem 2

Teacher: (*Display Overhead Modeling page of Review Problem 2. Have students look at Review Problem 2.*)
  "The Empire State Building is 1,472 feet tall. It is 1/20 as tall as Mount Everest. How tall is Mount Everest?"
  Touch Problem 2. (*Guide students to identify the problem type and solve Review Problem 2.*) What's the first step in FOPS?

Students: Find the problem type.

Teacher: Great. To find the problem type, what must you do first?

Students: Read the problem and retell it in your own words.

Teacher: Good. (*Read the problem and call on a student to retell it.*) Remember, when you retell a problem, you use your own words and tell what you know in the problem and what you are asked to find out.

Students: We know that the Empire State Building is 1,472 feet tall. We also know that the Empire State Building is 1/20 as tall as Mount Everest. We don't know how tall Mount Everest is, which we need to solve for.

Teacher: Great! You read and retold the problem. Now you need to ask yourself if this is an MC or a vary problem. What kind of a problem is this? How do you know?

Students: It's an MC problem, because it compares the height (in feet) of the Empire State Building to that of Mount Everest, and the compare words "1/20 as tall as" in the problem indicate a partial relation between the Empire State Building and Mount Everest.

Teacher: Now you are ready for Step 2: Organize the information in the problem using the MC diagram. To organize the information in your MC diagram, what must you do first?

Students: Underline the comparison sentence or question, circle the two things compared, and write them in the diagram.

Teacher: What is the comparison sentence or question in this story? How do you know?

Students: The second sentence, "It is 1/20 as tall as Mount Everest," is the comparison sentence, because the words "1/20 as tall as" tell about a partial relation between the compared and the referent.

Teacher: Underline this sentence as the comparison sentence. *(Pause for students to underline.)* What must you do next?

Students: Circle the two things compared in the comparison sentence.

Teacher: What are the two things compared in this problem? *(If students have difficulty, explain that "it" in the comparison sentence refers to the Empire State Building.)*

Students: The height of the Empire State Building and Mount Everest.

Teacher: From this comparison sentence, is the Empire State Building or Mount Everest the compared? How do you know?

Students: The Empire State Building is the compared, because it is compared to Mount Everest.

Teacher: Great. Write "Empire State" in your diagram for the *compared* and "Mount Everest" for the *referent*. From this comparison sentence, what is the *relation* between Empire State and Mount Everest in feet?

Students: The Empire State Building is 1/20 as tall as Mount Everest. So the relation is 1/20, which is a partial relation.

Teacher: Super! Write 1/20 for the *relation* in the diagram. *(Pause for students to complete.)* What do you do next?

Students: Read the problem to find the numbers given for the compared and referent and write them in the diagram.

Teacher: Great! Also, remember to write a question mark for the unknown information. *(Pause for students to figure out and check their work using the following completely mapped diagram.)*

$$\frac{\text{Empire State: 1,472 ft}}{\text{Mt. Everest: ? ft}} = \frac{1}{20}$$

Teacher: Now look at the diagram and read what it says. (*Pause for students to read.*) What must you solve for in this problem?

Students: The height (in feet) of Mount Everest.

Teacher: Good! You completed Step 2 using FOPS. What is the next step?

Students: Plan to solve the problem.

Teacher: Good. To solve this MC problem, you need to translate your diagram into a math equation. What is the math equation?

Students:

$$\frac{1{,}472}{?} = \frac{1}{20}$$

Teacher: Excellent. You completed Step 3 using FOPS. What is Step 4?

Students: Solve the problem.

Teacher: Good. What strategy can you use to solve for the question mark in the equation?

Students: Cross-multiplication.

Teacher: Good. Use cross-multiplication to solve the problem.

Students: ? × 1 = 1,472 × 20
? = 29,440

Teacher: What is the complete answer to this MC problem?

Students: 29,440 feet.

Teacher: Good. Write "29,440 feet" on the answer line and check your answer. (*Pause for students to write the answer.*) Does "29,440 feet" seem right? How do you know?

Students: Yes. The Empire State Building is 1/20 as tall as Mount Everest. That is, Mount Everest is 20 times as tall as the Empire State Building; that is, 20 × 1,472 = 29,440. So the answer "29,440 feet" is correct.

Teacher: Excellent work! Tomorrow you will practice solving more MC and Vary problems.

Lesson 12: Multiplicative Compare and Vary Review

Materials Needed

Checklists
: MC Problem–Solving Checklist and Vary Problem–Solving Checklist posters

  MC Problem–Solving Checklist and Vary Problem–Solving Checklist (laminated copies for students)

Diagrams
: Multiplicative Compare Problem and Vary Problem diagram posters

Student Pages
: Lesson 12: Review Worksheet 1

Teacher: (*Display posters of Multiplicative Compare Problem and Vary Problem diagrams and the two checklists in the classroom in a suitable place. Pass out Review Worksheet 1.*)

(*Review the two problem types—MC and vary—and the problem-solving steps for each problem type. Have students recall and distinguish the steps in the checklists, especially F—Finding the problem type and O—Organizing information using a diagram for the two different problem types. At this time, students should be able to readily recall the problem-solving steps without looking at their checklists. If they have difficulty remembering the steps, have them practice memorizing the steps as homework.*) I want you to do the four problems on this worksheet on your own. Remember to use the four steps to solve these problems.

(*Monitor students as they work. After some time, go over the answers. Make sure the diagrams are labeled correctly, the math equations are written and worked out correctly, and the complete answers are written on the answer lines; see below.*)

Review Worksheet 1, Problem 1: "In a spelling bee competition, Lee earned 45 points. Ben earned 2/3 as many points as Lee. How many points did Ben earn?"

$$\frac{\text{Ben: ? points}}{\text{Lee: 45 points}} = \frac{2}{3}$$

$$\frac{?}{45} = \frac{2}{3}$$

$$? \times 3 = 45 \times 2$$

$$? = (45 \times 2) \div 3 = 30$$

Answer: 30 points

"Use the data below to solve Review Worksheet 1, Problem 2."

Airplane	Capacity of Cargo Department
1	24 standard-sized bags

Review Worksheet 1, Problem 2: "Based on the capacity of each airplane, how many airplanes are needed to carry 72 standard-sized bags?"

(Explain the meaning of the word *capacity* to students before asking them to solve this problem. That is, the data in the table indicate that a total of 24 bags can fit into each airplane's cargo department.)

$$\text{If} \quad \frac{1 \text{ airplane}}{? \text{ airplanes}} = \frac{24 \text{ bags}}{72 \text{ bags}}$$
$$\text{Then}$$

$$\frac{1}{?} = \frac{24}{72}$$

$$? \times 24 = 1 \times 72$$

$$? = (1 \times 72) \div 24 = 3$$

Answer: 3 airplanes

Review Worksheet 1, Problem 3: "Richard washed 18 windows during the weekend. He washed 3 times as many as Ted. How many windows did Ted wash?"

$$\frac{\text{Richard: 18 windows}}{\text{Ted: ? windows}} = 3$$

$$\frac{18}{?} = \frac{3}{1}$$

$$? \times 3 = 18 \times 1$$

$$? = 18 \div 3 = 6$$

Answer: 6 windows

Review Worksheet 1, Problem 4: "Tara used 3 bags of candy bars to make 2 ice cream cakes. To make 6 ice cream cakes, how many bags of candy bars would she need?"

$$\text{If} \quad \frac{3 \text{ bags}}{? \text{ bags}} = \frac{2 \text{ cakes}}{6 \text{ cakes}}$$
Then

$$\frac{3}{?} = \frac{2}{6}$$

$$? \times 2 = 3 \times 6$$

$$? = (3 \times 6) \div 2 = 9$$

Answer: 9 bags

Lesson 13: Multiplicative Compare and Vary Review

Materials Needed

Checklists   MC Problem–Solving Checklist and Vary Problem–Solving Checklist posters

Diagrams   Multiplicative Compare Problem and Vary Problem diagram posters

Student Pages   Lesson 13: Review Worksheet 2

Teacher: (*Display Multiplicative Compare Problem and Vary Problem posters and checklists in the classroom in a suitable place. Pass out Review Worksheet 2.*)

Today, you will solve five problems on this worksheet on your own. Remember to use the four steps to solve these problems. (*Have students refer to their checklists only as needed.*)

(*Monitor students as they work. After some time, go over the answers. Make sure the diagrams are labeled correctly, the math sentences are written and worked out correctly, and the complete answers are written on the answer lines; see below.*)

"Use the table below to solve Review Worksheet 2, Problem 1."

Person	Weight (pounds)
David	180
Joshua	240
Danisha	150
Joe	140
Freya	60
Carol	?

Review Worksheet 2, Problem 1: "Joshua weighs how many times as many pounds as Freya?"

Joshua: 240 pounds
Freya: 60 pounds

$$\frac{240}{60} = ?$$

$$? = 240 \div 60 = 4$$

Answer: 4 times

Review Worksheet 2, Problem 2: "Ms. Rivera used 9 apples to make 6 apple pies. If she has only 3 apples, how many apple pies can she make?"

$$\text{If} \quad \frac{9 \text{ apples}}{3 \text{ apples}} = \frac{6 \text{ pies}}{? \text{ pies}}$$
Then

$$\frac{9}{3} = \frac{6}{?}$$

$$? \times 9 = 3 \times 6$$

$$? = (3 \times 6) \div 9 = 2$$

Answer: 2 pies

Review Worksheet 2, Problem 3: "Ms. Young planted 36 roses in her yard. She planted 2/3 as many roses as Mr. Shultz. How many roses did Mr. Shultz plant?"

$$\frac{\text{Young: 36 roses}}{\text{Shultz: ? roses}} = \frac{2}{3}$$

$$\frac{36}{?} = \frac{2}{3}$$

$$? \times 2 = 36 \times 3$$

$$? = (3 \times 36) \div 2 = 54$$

Answer: 54 roses

Review Worksheet 2, Problem 4: "Mr. Robins drives 21 miles every working day. Mr. King drives 4/7 as many miles as Mr. Robins. How many miles does Mr. King drive every working day?"

$$\frac{\text{King: ? miles}}{\text{Robins: 21 miles}} = \frac{4}{7}$$

$$\frac{?}{21} = \frac{4}{7}$$

$$? \times 7 = 21 \times 4$$

$$? = (21 \times 4) \div 7 = 12$$

Answer: 12 miles

"Use the data below to solve Review Worksheet 2, Problem 5."

Bagel Prices

Plain	2 for $1
Garlic	4 for $3
Cinnamon	3 for $2

Review Worksheet 2, Problem 5: "If Shapiro's Bagels sold 120 garlic bagels on Monday, how much money did the shop make that day on the garlic bagels."

$$\text{If } \frac{4 \text{ bagels}}{\text{Then } 120 \text{ bagels}} = \frac{\$3}{\$?}$$

$$\frac{4}{120} = \frac{3}{?}$$

$$? \times 4 = 120 \times 3$$

$$? = (120 \times 3) \div 4 = 90$$

Answer: $90

Teacher: Excellent work, everyone! Remember, to be good problem solvers, use the four steps on your problem-solving checklists. These steps will help you to solve many different multiplication, division, addition, and subtraction problems.

About the Author

Asha K. Jitendra, PhD, is a professor of special education within the College of Education at Lehigh University. She earned her doctorate in 1991 at the University of Oregon. Jitendra's research and teaching interests involve designing effective mathematics and reading instructional approaches for academically diverse learners, including students with disabilities, students who are at-risk, and students from culturally and linguistically diverse backgrounds. In addition, her research has focused on textbook analysis and aligning assessment and instructional practices with the ultimate goal of promoting access to the general education curriculum for students with disabilities. She has managed several federal research grants. Recently, she and her colleague, Jon Star, assistant professor at Michigan State University, were awarded the Mathematics and Science Research grant by the Institute of Education Sciences. The 3-year project will extend the previous work on schema-based instruction to teach mathematical problem solving to middle school students.

Jitendra's scholarly contributions include more than 60 publications in peer-reviewed journals that include conceptual, descriptive, and quantitative publications on a range of topics. Her publications have appeared in numerous journals (e.g., *Exceptional Children, School Psychology Review, Journal of Special Education, Journal of Learning Disabilities,* and *Journal of Educational Research*). She and her colleagues were recognized by the American Psychological Association with an award for an outstanding article in the *Journal of School Psychology.* Jitendra has presented nationally and internationally on effective instructional strategies for enhancing the academic performance of children with learning disabilities. She serves on seven editorial boards. In addition, she served as the associate editor of the *Journal of Learning Disabilities* and edited two special issues for other journals (i.e., Textbook Evaluation and Modifications for Students with Learning Problems for *Reading and Writing Quarterly* and Mathematics Assessment for *Assessment and Effective Intervention*).